2027 A GLOBAL CRUCIBLE

2027 A GLOBAL CRUCIBLE

K.J. STERLING

Contents

Dedicated to all of the service members stationed overseas, those ready to deploy, and the families that sacrifice to enable them to protect our country. God bless the United States of America! God bless us all!

Phase O

December 2026: When Time Stood Still.

The chill of December 2026 had descended upon the world, yet for many, it was a season of anticipated warmth, a collective pause before a storm that everyone felt brewing but no one dared name. In San Diego, the palm trees swayed in a gentle breeze, their fronds a familiar green against the pale winter sky, adorned now with strings of twinkling holiday lights. The air, crisp and clear, carried the scent of pine and cinnamon from a nearby street vendor, a pleasant change from the usual ocean salt.

Petty Officer Second Class Elena "Ellie" Rodriguez, usually confined to the sterile hum of the USS Ronald Reagan's intelligence center, found herself in Balboa Park, bundled against the unusual cold. Her sister, back in San Diego, had practically dragged her out, insistent on a moment of civilian normalcy before Ellie shipped out again. Laughter echoed from families strolling through the festive displays. Elena, normally analytical and reserved, felt a flicker of genuine peace as she watched children chasing pigeons near the Spreckels Organ Pavilion. It would not be long before the hum of the Reagan was her constant companion again. The news cycles were a cacophony of dire warnings and thinly veiled threats, but here, for a precious moment, it was just holiday cheer.

Thousands of miles away in Taiwan, Sergeant Kai Chen wiped sweat from his brow with the back of a gloved hand, the coarse fabric of his uniform scratching against his skin. The December air was humid and thick, clinging to the skin like a damp shroud, a stark contrast to the festive scenes he imagined back home in Pennsylvania. There was no festive cheer here, only the oppressive heat of the green jungle, and the invisible growing tension. Beijing's rhetoric had escalated dramatically in the last six months. He pulled out a crumpled photo from his breast pocket. It was his younger sister, graduating from high school, beaming. He wondered if they truly understood how close to the edge they felt out here.

To the north east, on Kadena Air Base Okinawa, Japan, Captain Anya "Viper" Sharma felt the pre-dawn chill bite at her exposed skin as she walked across the tarmac. Even in December, the island air carried a subtle hint of the tropics, but the wind off the East China Sea was sharp. Her beast, an F-35A Lightning, sat ahead, its stealthy, angular silhouette a predatory gleam against the dark sky. Today's mission, like so many lately, was a "presence patrol" in the disputed airspace over the Senkaku Islands, a diplomatic dance with real-world stakes. Chinese J-20s and H-6 bombers had become increasingly aggressive in their intercepts, testing boundaries, pushing envelopes.

She thought of her family in Texas. Her parents, bless them, had tried to convince her to pursue a safer, civilian career. But Anya had always been drawn to the sky, to the challenge, to the idea of defending something larger than herself. Now, that something larger felt acutely vulnerable. She climbed the ladder and settled into the cockpit, the familiar embrace of the G-suit tightening around her legs. The F-35 was more than just a fighter; it was a node in a vast, interconnected network, a single point in a vast web of information.

On the frozen edge of the Demilitarized Zone in South Korea, Specialist David "Dave" Miller pulled his fleece neck gaiter higher, trying to ward off the frigid air that seemed to cut through layers of thermal gear. His M-LOK rifle felt like an extension of his arm. North Korea, backed by its increasingly vocal patrons in Beijing and Moscow, had ramped up its provocations to an unprecedented level. Artillery drills pounded the northern side of the DMZ daily, their concussions sometimes rattling the very ground beneath Dave's boots.

"Yo, Miller, feel like a popsicle yet?" Private First Class Henderson, a wisecracking medic, stomped his feet to warm them. Dave grumbled, his voice muffled by the gaiter. "I just need a stick shoved" BOOM! Another artillery round goes off.

He thought of his little brother, still in high school, excited about joining the basketball team. To his mom, who worried constantly, her calls filled with thinly veiled questions. He always told her the same thing: "Everything's fine, Mom. Just another day in paradise". The lie felt heavier with each passing week. The old order, strained by two decades of simmering resentments and shifting allegiances, was finally cracking.

In the chill of late 2026, the cracks were visible, deep fissures running through the geopolitical landscape. China, its economic might now matched by a formidable military, flexed its muscles in the South China Sea, while North Korea, emboldened by its benefactors, bristled with renewed aggression against its southern neighbor. And then there was the Middle East, a perpetual tinderbox now ablaze on seven fronts for Israel, though none burned hotter than the endless, grinding conflict with Iran. A new axis was forming, Tehran, Moscow, Pyongyang and Beijing, bound by a shared ambition to dismantle the Western-led order, to rebalance a world they saw as skewed.

America, a colossus burdened by the weight of its commitments, stood by its allies and partners, a bulwark against the encroaching tide. The lines were drawn, not just on maps, but in the hearts and minds of soldiers, sailors, airmen, and marines who knew, with chilling certainty, that their time was coming. The fight would be a political nightmare, an economic strangulation, and, most terrifyingly, a bloodbath.

The world was on the precipice, teetering on the brink of a new, horrifying chapter. Public anxiety was at an all-time high, fueled by relentless media speculation and the escalating rhetoric from world leaders. Opinion polls in Western nations showed an acceptance of the inevitable conflict, a fatalistic dread mixed with a quiet determination.

Calls for diplomatic solutions were increasingly drowned out by demands for strength and preparedness. Despite the festive veneer of the holidays, an undercurrent of fatalistic dread permeated the season, a chilling certainty that the peace, like the fleeting warmth of a winter fire, was beautiful but ultimately ephemeral. The world was simply holding its breath.

Phase I

February 2027: The Gathering Storm

The year is 2027, and though the world is still full of the rituals of daily life, an undercurrent of unease ripples beneath the surface. Familiar alliances strain under the weight of unspoken grievances, and quiet escalations blur the line between diplomacy and brinkmanship. Governments speak in measured tones but their actions betray a growing apprehension. A sense that the global order is shifting, subtly but irreversibly. In this fragile calm, individuals across continents begin to sense a menacing tension coiling, ready to strike. Their choices, shaped by culture, loyalty, and circumstance, will soon intersect in ways none could foresee.

The Silent Watch Begins Again

CTT2 Elena "Ellie" Rodriguez, US Navy, USS Ronald Reagan

The hum of the Ronald Reagan's internal systems was Elena Rodriguez's constant companion, a deep, resonant thrum that vibrated through the deck plates and settled into her bones. Eighty feet below the flight deck, in the windowless, climate-controlled sanctum of the intelligence center, Ellie was just another cog in the vast, complex machine of the carrier strike group. Her station, a sleek console bathed in the cool glow of multi-spectral displays, hummed with a softer, more focused energy. As a Cryptologic Technician Technical, or CTT, her world was data – the unseen currents of electronic warfare, the whispers and shouts of an increasingly crowded electromagnetic spectrum.

The West Pacific was anything but passive. For weeks, the Reagan had been running high-tempo operations, a dance of power projection and deterrence in waters that felt less like international territory and more like a contested arena. Every ping, every burst transmission, every faint signature was a piece of a puzzle Ellie helped assemble.

Today, the puzzle was particularly challenging. Her fingers, quick and precise, navigated the multi-touch screens of her analysis suite. The system, code named "Argus," was cutting-edge, a quantum leap from the clunky interfaces she'd trained on just a few years prior. Argus, powered by advanced AI and machine learning algorithms, sifted through petabytes of intercepted signals, identifying anomalies, correlating seemingly disparate data points with chilling speed and accuracy. Most days, it was tedious, a digital fishing expedition yielding mostly commercial traffic or benign military exercises. But lately, Argus was chirping with increasing urgency.

"Anything spicy, Rod?" Lieutenant Jenkins, her OIC, leaned over her shoulder, his voice low over the ambient hum. He had the perpetually tired eyes of a man who hadn't seen the sun in weeks, a common affliction in the intel center. Ellie shook her head, but her gaze was fixed on a series of overlaid patterns on her primary display. "Not 'spicy,' sir. More... discordant. Like an orchestra tuning up, but some instruments are playing the wrong notes." She gestured to a cluster of signatures, usually indicative of a particular type of maritime patrol activity, but these were subtly different. Frequency shifts, unusual encryption methodologies, staggered transmission times. "It's what Argus flagged this morning. Not quite matching known profiles for these assets."

Jenkins squinted at the data. "Ghosting the P-8s again, or something new?"

"New. Or at least, new patterns. Not PRC, not Russian. Not exactly." Ellie zoomed in, Argus highlighting specific packets. "It's almost as if they're mimicking known signatures, but with a subtle deviation. Like a dialect, not a language."

This wasn't theoretical; it was the meat of her job. In the invisible war of electromagnetic waves, disguise was a potent weapon. A nation's electronic fingerprint was as unique as its true identity. And someone, somewhere out there, was trying to blur the lines. Her comms buzzard discreetly. It was a message from her sister, back in San Diego. A quick, terse update: "Mom's still worried. Saw the news about the Korean DMZ. Stay safe." Ellie typed a quick reply – "All good here. Tell Mom not to worry." – knowing it was a hollow platitude. Everyone was worried.

Life on the Reagan was a paradox: claustrophobic yet expansive. Thousands of souls confined within steel walls, yet connected to the vast, simmering world outside by an umbilical cord of data and responsibility. They were a floating city, a projection of American power, and a primary target should the unthinkable happen.

Later, during a rare lull, Ellie found herself in the ready room, catching a brief moment of quiet before the next watch. The constant thrum of the ship was a dull roar here, punctuated by the faint, rhythmic clang of something being secured above. She scrolled through news feeds on her personal tablet, the unclassified net a filtered window into the chaos. Ukraine, still brutal. Israel, still besieged. And always, the quiet escalation in the Pacific, a war of words and veiled maneuvers that felt increasingly like a prelude.

"You look like you're carrying the weight of the world, Rod." AB1 Marcus "Rhino" Evans, a burly aviation boatswain's mate from the flight deck crew, dropped into the seat opposite her. He smelled faintly of jet fuel and sweat, a reminder of the raw power above them. Ellie managed a weak smile. "Just the usual data deluge. Sometimes it feels like all the bad news from every corner of the globe converges right here."

Rhino nodded, his face serious. "It ain't just data anymore, is it? We're hearing talk on the deck. Readiness levels. Sorties running hotter. Everyone feels it. Like the whole world is just waiting for someone to light the fuse." He paused, then leaned forward. "You think it's coming, Rod? The big one?" Ellie looked out at the steel bulkheads, the rows of empty chairs. She thought of the discordant signals, the mimicking frequencies, the subtle shifts in the electronic landscape. She thought of Argus's increasingly frequent red flags. "I think," she said slowly, "we're already in it! We just haven't started shooting yet."

The thought hung heavy in the air, a truth they both instinctively understood. On the USS Ronald Reagan, in the vast, unforgiving expanse of the West Pacific, the silent watch continued.

And with every passing day, the silence grew heavier, more ominous, laden with the unspoken promise of the storm to come.

Living On The Island Of Interest

Sergeant Kai Chen, US Marine Corp, Taipei, Taiwan

The air in Taiwan in February was humid and thick as always. For Sergeant Kai Chen, a squad leader with the 3rd Marine Division, the tropical heat was a constant, unwelcome companion, a stark contrast to the arid landscapes of his previous deployments. His M41 carbine, a sleeker, modular evolution of the M4, felt solid and familiar in his grip, its composite frame cool despite the warmth. His squad was dug in, as they had been for the past three weeks, on a ridge overlooking a fertile valley just outside Taipei.

Their position was a meticulously constructed fighting hole, reinforced with sandbags and camouflaged with local vegetation. It wasn't a trench line, not yet, but it felt like the beginning of one. The tension was palpable, a live wire humming beneath the surface of their mundane existence. Every rustle in the dense foliage, every distant drone of a civilian aircraft, sent a ripple of heightened alert through the squad. They were here as part of a "joint training exercise" with the Taiwan Armed Forces, but everyone knew the real reason. Threats giving way to thinly disguised ultimatums. Naval drills in the Taiwan Strait had become almost daily occurrences, their scale and aggression increasing with each passing week.

"Sarge, you see that?" Lance Corporal Miller, a fresh-faced twenty-year-old with an almost unnerving enthusiasm, pointed a gloved finger towards the hazy horizon. His enhanced optic goggles, part of the new integrated combat system, provided crystal-clear magnification. Kai followed his gaze. A faint smudge against the pale blue sky. "Looks like another 'commercial' flight deviating from its usual air corridor." The sarcasm in his voice was thick. These "commercial" flights, often large cargo planes or even repurposed civilian jets, had a habit of buzzing Taiwan's ADIZ, testing response times, probing vulnerabilities. This was psychological warfare, a constant reminder of the sword hanging over the island.

"Think they'll try anything today?" Miller asked, his voice betraying a hint of anticipation, a youthful eagerness for action that Kai had long since outgrown.

"They'll try to get under our skin, Miller. Same as yesterday. Same as tomorrow." Kai adjusted his position, his knees protesting slightly. The 'futuristic' touches of their gear were impressive – the integrated comms in their helmets, the smart scopes that could identify targets and provide real-time ballistic solutions. Some of the heavier weapons teams utilized powered load-bearing exoskeletons, designed to augment strength and endurance without significantly impeding mobility. But at the end of the day, it was still dirt, sweat, and waiting. And the waiting was the hardest part.

He ran a diagnostic on his helmet's tactical display. The feed showed the positions of his squad members, overlaid with topographical data and the occasional blip from a distant drone. The drones were their eyes in the sky, silent sentinels, but they couldn't see everything. The jungle was a master of concealment.

His thoughts drifted to his family back in Pennsylvania. His parents, still holding onto the immigrant dream, had tried to convince him to choose a safer path. But Kai had felt the pull of service, a need to defend the ideals he believed in. Now, those ideals felt very tangible, and very much under threat, here on this humid island. He pulled out a crumpled photo from his breast pocket – his younger sister, graduating from high school, beaming. He missed her easy laughter, the quiet comfort of home. He wondered if they truly understood how close the edge felt out here. The news filtered back to them, sanitized and spun. Here, it was raw, visceral.

A whisper of wind rustled through the palm fronds, a sound that could be anything or nothing. Kai scanned the tree line, his eyes trained, his senses heightened. They had been conducting joint patrols with Taiwan units, an exercise in interoperability that felt more like a dress rehearsal. The Taiwan soldiers were disciplined, professional, but beneath their stoicism, he sensed the same gnawing apprehension. This was their home, their sovereignty on the line.

"Sarge, comms traffic just spiked," PFC Matthews, the squad's comms specialist, murmured, her voice tight. "Heavy encryption. Coming from the Strait."

Kai's gaze sharpened. "Any patterns?"

"Not yet. Just... a lot of it. More than usual for a 'drill.'"

He stood up slowly, keeping low, peering through the dense canopy. The air felt heavier, thicker, as if the rising tension had a physical weight. The "silent watch" that Ellie was keeping thousands of miles away felt profoundly connected to the "green hell" Kai found himself in. The world was holding its breath, and Taiwan, he knew, was very much at the center of that tension.

He checked his M41 again, then glanced at his squad, their faces grim, their eyes scanning the treeline. They were ready. Or as ready as they could be for something no one truly wanted. The waiting continued, a slow, agonizing crawl towards an inevitable dawn.

The war they don't show you isn't silent. It just hasn't echod yet.

Phase II

March 2027: A Deterrent Or A Provocation

The fragile veneer of peace, already thin, began to fray more rapidly in the early months of 2027. Across the vast, shimmering expanse of the South China Sea, Beijing, with a chilling blend of calculated aggression and unveiled ambition, escalated its naval operations. The sleek, formidable gray hulls of Chinese warships, their anti-ship missiles poised, cut through the azure waters with increasing frequency. They weren't just exercises; they were assertions of dominance, relentless, aggressive drills that pushed ever closer to Taiwan's nervous shores, each maneuver a silent, yet deafening, declaration of intent. The world watched, transfixed, as China, like a coiled dragon, flexed its formidable military might.

Simultaneously, in the unseen war that simmered beneath the surface of the interconnected world, a shadow war unfolded. Chinese state-backed hackers, a silent army of digital warriors, probed and prodded at the very heart of American defense. Like phantom fingers, they reached into U.S. military networks, testing vulnerabilities, mapping weaknesses, and laying dormant traps for a conflict that felt increasingly inevitable. The digital echoes of their intrusions resonated in secure intelligence centers, setting off alarms that few outside those hallowed walls fully comprehended.

The last bastions of diplomacy crumbled. High-level talks between Washington and Beijing, once heralded as a beacon of hope, collapsed into recriminations and accusations. Each side, entrenched in its own narrative of righteousness, pointed fingers at the other, condemning them for destabilizing a region already teetering on the brink. The air grew thick with mistrust, the carefully constructed bridges of communication consumed by the flames of mutual suspicion.

And while the Pacific bristled with heightened tension, the frozen expanse of Eastern Europe continued to bleed. Russia, a bear awakened from a long slumber, tightened its grip on Ukraine, pushing deeper into contested territories with a brutal, relentless resolve. Tanks rolled, artillery pounded, and the cries of a struggling nation were lost in the roar of a burgeoning war machine. The echoes of that distant conflict reverberated globally, mirroring the aggressive drills near Taiwan and the chilling tests of long-range missiles by North Korea, each act a synchronized beat in a macabre global symphony.

The economic arteries of the world, already strained, began to seize. The U.S. and its allies, in a desperate attempt to staunch the flow of aggression, unleashed a barrage of severe sanctions against China and Russia. It was an act of economic warfare, a gamble with global stability, and the retaliation was swift and brutal. Cyberattacks, far more sophisticated and widespread than any seen before, crashed upon Western financial institutions, disrupting markets, siphoning funds, and sowing widespread panic. The digital assault was a chilling precursor of the chaos to come.

Yet, even amidst the escalating global tensions, the Middle East remained its own volatile crucible. Iran, defiant and unyielding, accelerated its nuclear program, its centrifuges spinning ever faster, producing a chilling threat to regional stability. This brazen defiance pushed Israel, a nation accustomed to fighting for its very existence, to a desperate brink. In a series of precision strikes, Israel conducted preemptive attacks on key Iranian nuclear facilities, sending shockwaves through an already fractured region. The retaliatory threats from Tehran were immediate and apocalyptic, adding yet another volatile ingredient to the simmering global cauldron. The world was suspended in the agonizing silence before the storm. Everyone knew it was coming and no one knew the right way to stop it.

The Sky Sentinel

Captain Anya "Viper" Sharma, US Air Force
Kadena Air Base, Okinawa, Japan

Anya ran a hand over the composite skin of her jet. It felt cold, inanimate, yet she knew every inch of it, every system, every nuance of its power. The F-35 was more than just a fighter. It's a flying supercomputer capable of fusing data from countless sources and presenting an unparalleled picture of the battles pace. Futuristic, yes, but its core purpose was ancient: to dominate the skies.

"Morning, Ace," Sergeant Jenkins, her Maintenance Manager, jokingly greeted her with a nod.. He was already performing his final preflight checks, a familiar, rhythmic ritual of taps and visual inspections.

"Morning, Jenkins. She purring for me?" Anya asked, while pulling on her flight gloves.

"Like a baby dragon, Cap. All green. Fuel, weapons, comms, stealth profile, everything's dialed in."

Anya climbed the ladder and settled into the cockpit, the familiar embrace of the G-suit tightening around her legs. The cockpit itself was a marvel, dominated by large, customizable multi-function displays replacing traditional dials, offering a fluid, panoramic interface. Her helmet, a cutting-edge piece of augmented reality tech, would overlay flight data, target cueing, and real-time sensor information directly onto her field of view, effectively projecting critical data onto the canopy itself. It was like having X-ray vision and a tactical map projected onto the world itself.

As she powered up, the F-35's systems hummed to life, a symphony of whirs and clicks. The data links snapped open, connecting her to the larger network – to command and control, to AWACS, to other flights in the air, and theoretically, even to ground units and naval assets. She was a single point in a vast web of information.

"Tower, Viper Three-Two, ready for taxi," she intoned into the mic in her mask.

The controller's voice crackled back. "Viper Three-Two, cleared to taxi runway zero-five. Winds zero-eight-zero at ten, gusting fifteen. Report at hold short."

As she taxied out, the weight of her mission pressed down on her. The news had been grim even here on base. Rumors of increased deployments to South Korea, the sheer scale of the Chinese naval build-up, and the chilling reports from Ukraine, where Russian air superiority was still heavily contested, served as a constant reminder of the stakes.

China seemed intent on proving their technological parity, or even superiority, in some areas here too. She did a final systems check, her eyes sweeping across the large cockpit displays. The targeting pod showed the crisp outlines of distant islands, the radar painting an empty sky. For now.

"Viper Three-Two, cleared for takeoff."

Anya pushed the throttle forward, the immense power of the F-35's engine pressing her back into her seat. The jet surged down the runway, gathering speed with brutal efficiency. In moments, the wheels lifted, and she was airborne, climbing rapidly into the predawn darkness. Below, the lights of Okinawa twinkled, a fragile tapestry against the vast ocean. Above, the stars were cold and indifferent.

She was the sentinel, one of many, guarding the fragile peace of the Pacific. But Anya knew, with every fiber of her being, that peace was a brittle thing, and the sky, once a boundless expanse, was shrinking, becoming a contested domain where extensive operations displaying force, pushing boundaries, and potentially where the next global conflict could begin. She gripped the stick, her eyes scanning the horizon, ready for the inevitable.

THE FROZEN WAR
THAT NEVER ENDED
SPECIALIST DAVID MILLER
US ARMY
CAMP CASEY, SOUTH KOREA,
NEAR THE DMZ

The Frozen War That Never Ended

Specialist David "Dave" Miller, US Army
Camp Casey, South Korea, near the DMZ

The biting wind of a South Korean early spring morning was a physical assault. Not as bad as January but still much colder than any day back home in Tennessee. For Specialist David Miller, an Army infantryman with the 2nd Infantry Division his M-LOK rifle, equipped with a smart optic that could project target data onto his eye-piece, felt like an extension of his arm. It was cold but the tension along the Demilitarized Zone (DMZ) was even colder.

They were on patrol, a routine sweep of their sector just south of the DMZ, but nothing felt routine anymore. The air crackled with a low-frequency hum of unease. North Korea, backed by its increasingly vocal patrons in Beijing and Moscow, had ramped up its provocations to an unprecedented level. Infiltrations, though rare, were becoming bolder, and the propaganda blared incessantly from colossal loudspeakers across the border, a constant psychological assault.

"Yo, Miller, feel like a popsicle yet?" Hendo never got tired of this joke. "Almost there, Hendo. Just need a cherry-flavored stick," Dave replied. He scanned the dense, leafless forest, every shadow a potential threat. The new tech added to their gear was subtle but significant: the advanced night vision that made the darkest nights seem like twilight, the integrated GPS and mapping systems in their wrist-mounted comms units, and the highly sensitive acoustic and thermal sensors that could detect even faint traces of human movement or heat signatures. Yet, it was still a foot patrol, still slogging through mud and thickets.

Their patrol route took them past reinforced checkpoints, past concrete bunkers scarred with the marks of previous, smaller skirmishes. This wasn't a static line; it was a volatile, unpredictable frontier. The DMZ itself, an almost surreal strip of pristine wilderness, was also the most heavily fortified border on Earth. And they were right on its frozen edge.

They paused at a designated observation point, a small, camouflaged bunker overlooking a section of the wire. Dave pulled out his optics, a pair of next-generation binoculars that could zoom in with incredible clarity and analyze thermal signatures. He swept the Northern side, the desolate, grey landscape stark against the sky. He saw the usual: guard towers, faint trails, the occasional flicker of movement. And then, something else.

"Hendo, tell me what you see at eleven o'clock, about two klicks out," Dave murmured, handing him the binoculars. "Looks like... new construction?" Henderson peered through them for a long moment. "Damn. You're right. Looks like they're reinforcing that old artillery position. And... are those new sensor arrays? "Dave nodded grimly. "Looks like it. They're digging in. And they're not hiding it anymore."

The implication hung heavy. This wasn't just about provocations; it was about preparation. North Korea, with its increasingly advanced missile and artillery capabilities, was a direct and immediate threat to Seoul, a sprawling metropolis just miles south of the border. And behind North Korea stood Russia and China, supplying and emboldening.

A wave of dread, sharp as the cold wind, washed over Dave. He had joined the Army for a sense of purpose, for adventure, for something beyond his small hometown. Now, that purpose felt crushingly real, and the adventure terrifyingly imminent. He thought of the endless back-and-forth between Russia and

Ukraine, the grinding attrition, the human cost. He thought of Israel's multi-front war. And now, this, on the Korean peninsula. He looked at his squad, their faces etched with the same weariness, the same silent understanding. They were young, but the world had aged them. They were ready, as ready as any soldier could be. But the fight, he knew, would be unlike anything humanity had seen in generations. On the frozen edge of the DMZ, Specialist David Miller waited, dreading the dawn of a world at war.

Phase III

X-Hour D-Day: The Fire Ignites

The dawn of April 17th, 2027, did not break; it shattered. It was not painted in the gentle blush of a new day, but in the searing, incandescent flash of a thousand simultaneous explosions, a global sunrise of war. The coordinated offensive was a masterstroke of timing and deception, striking like a hydra from the heart of the Pacific to the frozen edge of the Korean Peninsula.

At 07:30 EST, from the hallowed halls of the White House, the US President's voice, strained but resolute, cut through the static of fear on an emergency broadcast: "Today, the forces of aggression have launched an unprovoked and coordinated attack against our Military forces, our allies and interests in the Pacific and on the Korean Peninsula. Let there be no doubt: the United States will respond with overwhelming force, standing shoulder to shoulder with our partners. This aggression will not stand. We are at war." The words, stark and undeniable, echoed across a stunned nation.

At 09:00 EST, Chairman of the Joint Chiefs of Staff, stood before a phalanx of hungry reporters at the Pentagon, his face etched with grim determination. "This is a full-spectrum attack," he stated, his voice a low rumble, "unprecedented in its coordination and technological sophistication. Our forces are engaged across multiple theaters. We anticipate heavy fighting and significant challenges, but our resolve is unwavering." The words hung heavy in the air, a stark prophecy of the bloodshed that had just begun. The world, caught in the inferno, had no choice but to burn.

16:00 CST, Beijing's response, delivered with an icy calm by a spokesperson for the Ministry of National Defense, was broadcast to the world: "In response to decades of hostile provocation and attempts to undermine the sovereignty and security of the People's Republic of China, and to protect the legitimate interests of our strategic partners, defensive military operations have commenced in the region. We urge all foreign forces to cease interference and withdraw immediately." It was a declaration in a false veil of self righteousness, but its meaning was chillingly clear.

Across the Middle East, the voice of the Supreme Leader of Iran, resonated with triumphant fire on state television: "The Great Satan and its Zionist puppets now reap what they have sown. The era of oppression is over. The forces of true justice are rising across the world, united in their righteous cause." His words, a poisonous balm to his followers, were a declaration of Jihad, a holy war, against a world he despised.

Panic, a raw, primal scream, tore through global markets. Stock exchanges, once the beating heart of world commerce, convulsed and flat lined, trading suspended as major indices plummeted into free fall. News channels, initially scrambling in bewildered disbelief, quickly became the grim-faced heralds of chaos, their round-the-clock coverage a relentless stream of unfolding horrors. Grim-faced military analysts, their voices devoid of pretense, and solemn government officials urged a calm that felt like a cruel joke.

Public sentiment, at first paralyzed by shock and disbelief, quickly curdled into raw, visceral fear as the dizzying scale of the conflict became terrifyingly apparent. Protests, which had simmered for months like an angry fever, now erupted into full-blown riots in several European capitals, only to be violently dispersed by riot police, their truncheons swinging like pendulums of despair.

The World At War

CTT2 Ellie Rodriguez, USS Ronald Reagan, West Pacific

The Ronald Reagan's intelligence center, usually a sanctuary of controlled quiet, erupted into a maelstrom of flashing lights and piercing alarms. Ellie was at her station, just finishing her mid-watch, when Argus screamed. Not a chirping urgency, but a full-spectrum, system-wide wail that vibrated through her very teeth.

"Multiple, simultaneous frequency excursions!" a voice shouted from across the room. "Massive data spikes! Full spectrum jamming in the Taiwan and Luzon Straits!"

Ellie's fingers flew across her multi-touch console. Argus, designed to identify anomalies, was now overloaded with them. The discordant tuning she'd noticed months ago had resolved into a cacophony of coordinated electronic attack. Chinese naval and air forces were unleashing a devastating wave of cyber and electronic warfare, targeting satellite up-links, GPS signals, and encrypted comms across the Pacific.

"They're saturating the counter-jamming arrays, sir!" she yelled to Lieutenant Jenkins, pointing to a cascade of red indicators on her display. The enemy's EW capabilities were far more advanced than intelligence estimates had predicted, designed to overwhelm and degrade. "They're spoofing targeting data! Argus is showing ghost contacts everywhere!"

The ship lurched violently. Not a gentle swell, but a sharp, sickening roll that sent coffee mugs crashing. The air crackled with the stench of ozone and something acrid, smoke! Battle stations klaxons blared, an ear-splitting wail that cut through the electronic din. "Damage control reports! Section four, port side! We've taken a hit!" came a frantic voice over the ship's internal comms. Panic, cold and sharp, cut through the professional focus. A hit? Already? The Reagan was designed to withstand a beating, but this was too fast, too aggressive.

"Sir, Argus is confirming ballistic missile launches from mainland China!" Ellie shouted, her voice hoarse. Her hands trembled, not from fear, but from the sheer volume of data, the scale of the attack. The map on her main display was a horrifying tableau of inbound trajectories, painting a death-dealing web across the Pacific. "Multiple targets, sir! Targeting Guam, Japan, and naval assets!"

The ship shuddered again, a metallic shriek tearing through the hull. This wasn't a drill. This was it. Ellie gripped the edge of her console, her eyes fixed on the chaotic display, the silent war of signals now translating into very real, very violent explosions. The invisible became horrifyingly tangible.

The Green Hell Becomes A Red Inferno

Sergeant Kai Chen, Taipei, Taiwan

The day started the same as it did for many in this newly kinetic and chaotic environment. The first sound was a low, distant rumble, like an angry god clearing his throat. Then, the sky over the Taiwan Strait exploded. It wasn't one explosion, but a rapid, thunderous succession of them, artillery, rockets, and cruise missiles saturating the coastline! Sergeant Kai Chen, already in his fox hole, felt the ground beneath him buck and shudder with each concussive blast.

"INCOMING! BRACE FOR IMPACT!" his platoon sergeant roared, his voice barely audible over the growing roar.

The air was suddenly filled with the shriek of incoming rounds. The sounds were different from training; they were sharper, hungrier. Trees exploded into splinters. Dirt and rock rained down. Kai pulled his helmet lower, pressing himself deeper into the earth, the smell of cordite and pulverized concrete filling his nostrils.

His helmet's comms crackled. "Alpha Team, this is Command! Amphibious landing craft inbound! Massive force! Expect airborne drops! Engage! Engage! ENGAGE!"

"Miller! Get eyes on the beach!" Kai yelled, pushing himself up, peering through the scope of his M41. His smart optic was already struggling, flickering with jamming interference, but he could make out the dark shapes emerging from the hazy pre-dawn sea – an armada of landing craft, swarms of hovercraft, and low-profile assault vehicles surging towards the shore. Overhead, transport drones and tilt-rotor aircraft, previously unseen, materialized out of the low clouds, disgorging waves of troops. "They're everywhere, Sarge!" Miller's voice was a high-pitched gasp. "Thousands of 'em!"

"Light 'em up!" Kai barked, his voice raw. He squeezed the trigger of his M41, the rifle chattering rhythmically, spitting rounds into the advancing horde. Tracer fire crisscrossed the air, a deadly, beautiful dance. The integrated combat system in his helmet was feeding him target solutions, highlighting enemy combatants, but the sheer volume was overwhelming.

A barrage of heavy explosive rounds tore through the air near them, disintegrating a sandbag barrier and throwing up a geyser of earth. These weren't just massed infantry; they were equipped, relentless, and coming hard. "Heavy contact, front and left flank!" Matthews, the comms specialist, screamed. "Sarge, we're taking heavy casualties on the next ridge! They're pushing hard!" Kai saw it then – the first wave of enemy soldiers, clad in dark, advanced combat suits, swarming up the incline, their rifles spitting fire. This wasn't a drill or a provocation. This was a full-scale invasion. The green hell had become a red inferno.

Deterrence No Longer! Cleared HOT!

Captain Anya "Viper" Sharma, Kadena Air Base, Okinawa

The scramble alarm had torn Anya from a restless sleep at 0430. But this wasn't the usual exercise. The siren was raw, unceasing, backed by the urgent shouts of ground crew. The Ops brief was terse, delivered by an angry frantic Colonel: "Coordinated kinetic and EW attack across the Pacific. Taiwan is under full-scale invasion. Korea is compromised. You are cleared hot for immediate engagement. Protect the skies."

Anya rapidly prepped and ran to her cockpit, the canopy slamming shut with a final, metallic thud. Her helmet-mounted display, usually a crisp, orderly interface, was a chaotic mess of overlapping alerts. The data links were struggling under heavy jamming, flashing "DEGRADED" warnings.

"Viper Three-Two, you are cleared for takeoff!" the tower screamed, their voice hoarse.

She launched into a sky that was no longer just dark, but alive with the terrifying glint of inbound missiles and the furious contrails of defending interceptors. Her F-35 roared off the runway, climbing sharply, pushing her deep into her seat.

"Multiple unidentified contacts, bearing zero-four-zero, fast-movers!" her flight lead's voice crackled, distorted by the jamming. "Looks like J-20s, maybe some newer variants. Engage! Engage! Engage!"

Anya's panoramic display showed the battle space lighting up like a Christmas tree, but a deadly one. Red triangles, indicating enemy aircraft, materialized and disappeared as her stealth systems and jammers fought against their sensors. The air was a ballet of electronic warfare, a dance of deception and counter deception.

"Viper Three-Two, target acquired! Bandit, one o'clock high, eighty miles, closing fast!" her integrated AI flight assistant, a sophisticated system, calmly informed her, despite the chaos. Anya's eyes locked onto the targeting cue in her helmet. A Chinese J-20, sleek and menacing, was vectoring hard towards her flight. It was fast, agile, and it was firing. The missile launch warning blared in her ears. "Flares and chaff! Break left! Hard!" she shouted, wrenching the stick. The F-35 responded instantly, pulling a gut-wrenching G-turn. Streaks of light blossomed behind her as countermeasures deployed. The enemy missile, guided by its advanced seeker, corrected, but her evasive maneuvers were just enough. It zipped past, a near-miss that could have been a quick end for Anya.

This wasn't an exercise. This was real. The sky, her domain, was now a killing field. She flipped her weapon systems to 'hot,' her finger hovering over the fire button. She could hear the strained breaths of her flight lead, the crackle of distant explosions. The "Sky Sentinel" is no longer observing; she was fighting, for her life, for her allies, for the very idea of a free Pacific.

A Frozen War's Rapid Thaw

Specialist Dave Miller, South Korea, DMZ

The first sound Dave heard on April 17th, 2027, was the wail of air raid sirens, immediately followed by the earth-shattering roar of a thousand North Korean artillery pieces. It was louder, more sustained, and infinitely more terrifying than any drill. The sheer volume of fire was unimaginable, a deluge of steel raining down on Camp Casey and the surrounding areas.

"GET TO THE BUNKERS! NOW!" a sergeant screamed, his voice swallowed by the thunder.

Dave, still shaking off the last vestiges of sleep, scrambled out of his barracks. The air was already thick with dust and the potent smell of high explosives. He saw men running, falling, the terrifying realization of what was happening etched on every face. This was it!

"Miller! Fall in! Fall in! We're moving to the forward positions!" Sergeant Riley yelled, pointing towards the sounds of the heaviest barrage.

The ground vibrated incessantly, a constant tremor that made it hard to stand. Buildings disintegrated into rubble. The sky, once serene, was now crisscrossed with the fiery arcs of North Korean rockets and the desperate, intercepting trails of South Korean and American air defense systems. But there were too many.

His wrist-mounted comms unit flashed. "URGENT BROAD-CAST: DPRK forces crossing DMZ along multiple sectors. Significant armor and infantry pushes detected. Chemical agents possible. Full MOPP level 4 imminent." Chemical agents. The words sent a fresh wave of ice through Dave's veins. He fumbled with his MOPP suit, his fingers numb with shock and adrenaline.

"CONTACT! NORTHERN WIRE BREACHED!" A voice screamed from their comms. Dave peered through the smoke and falling debris. Through the smart optic of his M-LOK, he saw them – hordes of North Korean soldiers, not just regular infantry, but what looked like heavily armored assault units, crossing the DMZ, swarming over the barriers they had observed for months. Some moved with an unnatural speed, perhaps assisted by lightweight powered assistance frames, their weapons spitting streams of conventional fire. Russian-made battle tanks, never before seen in such numbers, rumbled across the terrain, their heavy cannons already firing.

"Hold the line, boys! They want Seoul! We won't let 'em have it!" Riley roared, already laying down suppressive fire. Dave dropped to a knee, adjusting his optic. The system, miraculously still functioning, highlighted targets. He squeezed the trigger, the rifle kicking against his shoulder, sending bursts of rounds into the oncoming tide. He saw men fall, on both sides. The familiar faces of his comrades, moments ago joking, were now grim masks of determination and fear. This was no longer the frozen edge. This was the boiling point. The bloodbath had begun.

APRIL 17, 2027

The initial shock wave of the coordinated attack began to recede, replaced by the grinding reality of sustained combat. The meticulously planned offensive by the axis had achieved a devastating level of surprise and initial impact, forcing the Western allies into a desperate, reactive stance.

Fighting Through The Fog Of War

CTT2 Ellie Rodriguez, USS Ronald Reagan, West Pacific

The internal comms of the Reagan were a mix of clipped commands, frantic damage reports, and the controlled urgency of firefighting teams. The ship, though wounded, was still largely operational. The hit to Section Four had been a near-miss for the core systems, a lucky deflection, but the damage control teams were fighting rapidly spreading fires and containing ruptured lines.

In the intelligence center, the controlled chaos had solidified into stark efficacy. Argus, having weathered the initial EW storm, was now working overtime, trying to re-establish stable links and process the overwhelming flood of battle data. Ellie's console glowed, her fingers dancing across the screens, attempting to peel back the layers of enemy deception.

"We're getting fragmented reports, sir," Ellie reported to a sweat-drenched Lieutenant Jenkins. "Satellite imagery shows massed armor movements in North Korea, far beyond what intelligence projected. They've bypassed key choke points with vertical assault drones. And China... their air assets are pushing deep into our contested zones, faster than predicted."

"Any pattern to the EW attacks, Rod?" Jenkins asked, wiping a smear of grease from his cheek. "Are they still spoofing, or have they committed their full spectrum?"

"Still seeing adaptive mimicry, sir, but with a new layer," Ellie said, zooming in on a particularly complex cluster of signals. "It's not just spoofing. It's almost... predictive. Like their AI is anticipating our counter-measures and adjusting their signature before we deploy. We're losing visibility on key friendly assets, sir. Argus is showing half the task force as 'signal degraded' or 'unreliable.'"

The implications were chilling. If they couldn't trust their own sensor data, if their communications were compromised, then their cutting-edge force was fighting blind. The invisible war was as deadly as the kinetic one. Suddenly, a new alert flashed across her console – a desperate, garbled transmission from a distant P-8 maritime patrol aircraft. Argus worked to decode it, the words appearing slowly, agonizingly, on a smaller screen. "...taken heavy fire... anti-air missiles... multiple inbound... going dark..." Then silence. A terrible, crushing silence.

"Lost the P-8, sir," Ellie said, her voice flat, devoid of emotion, though her stomach twisted. That was a direct attack on an ISR asset, deep in what should have been their controlled space near the Sea of Japan. The enemy was bold, and their reach was terrifyingly long. Jenkins slammed his fist softly on the console. "Damn it. Get a trace. Anything. We need to know where that came from. We need to know what hit us."

Ellie nodded, her eyes narrowed, focused. The ship shuddered again, the sounds of distant explosions rattling through the hull. Survival depended on her ability to see through the electronic fog of war. Her hands flew, a blur of motion, battling an enemy she couldn't see, in a fight that felt like it was happening in the very air she breathed.

D+1

April 18th, 2027: A New Reality

The first full day of the conflict brought with it a menacing understanding of the enemy's capabilities and the sheer scale of their ambition. The initial shock gave way to the brutal rhythm of total war.

AI VS AI - A New Age Of Battle

CTT2 Ellie Rodriguez, USS Ronald Reagan, West Pacific

The intel center was a pressure cooker. Thirty-six hours had passed since the first strikes, and the air was thick with the smell of stale coffee and desperation. The Reagan was still running, a testament to its engineering and the damage control teams, but the persistent, adaptive electronic warfare attacks were grinding them down. Argus was now fully engaged in counter-EW, but it was a continuous, brutal battle.

"They've shifted spectrum again!" Ellie shouted, pointing to a new series of fluctuating signatures on her screen. "Their AI's learning faster than ours, sir! They're not just mimicking; they're evolving their jamming patterns based on our counter-efforts. We're losing situational awareness on the forward screens."

Lieutenant Jenkins rubbed his temples. "Casualty estimates from the Japan theater are coming in. It's bad, Rod. Very bad. And the P-8 that went dark... passive acoustic sensors picked up signatures consistent with an advanced submersible missile launcher, not a conventional sub. They've got assets we didn't account for." The implications hit Ellie like a physical blow. Not just air and ground, but deep undersea assets capable of launching anti-air missiles with devastating effect. The Chinese had achieved a level of naval stealth and long-range precision previously thought years away.

"Sir, new comms intercepts from the South China Sea. Fragmented, but consistent with... a significant naval engagement. Our Seventh Fleet is taking heavy contact." Ellie's fingers flew, trying to stitch together the broken pieces of data. The scale of the war was truly global, stretching resources thin, making every loss felt keenly.

The ship shuddered again, less violently this time, but a deep, resonant clang that indicated ordnance impacting the water nearby. Defensive systems immediately engaged, a series of rapid-fire thumps. This was the new normal – constant threats, constant vigilance, fighting an enemy that seemed to anticipate their every move.

The Assault Of An Endless Gauntlet

Sergeant Kai Chen, Taipei, Taiwan

The retreat was agonizing. Every kilometer inland was paid for in blood. Kai's squad, now a mere handful of exhausted, traumatized men, had fallen back from Taipei's outskirts, trading the dense jungle for the desperate, house-to-house fighting in a sprawling urban environment. The streets were choked with rubble and smoke, a thick cloud of the dust from the pulverized concrete and the metallic tang of fresh blood in our mouths.

"Clear!" Kai yelled, kicking open a shattered door, his M41 sweeping the dark interior of what was once a shop. Civilian casualties were everywhere – a horrifying testament to the speed and brutality of the invasion. There was no time to mourn, only to move, to fight. The Chinese forces were relentless. Their heavy infantry, aided by their powered load-bearing exoskeletons, moved with surprising speed through the urban maze, their advanced rifles rapidly spitting rounds. Small, agile attack drones, some equipped with thermal optics and even rudimentary facial recognition, zipped through the alleys, acting as mobile sentinels, forcing the Marines to constantly change position.

"Sarge, contact, second floor, building across the street!" PFC King, their new comms specialist after Matthews' death, whispered, his voice hoarse with fatigue. "Thermal signature. Three hostiles." Kai acknowledged with a grunt. He tossed a sensor grenade – a small, spherical device that emitted a burst of infrared and acoustic pings, mapping the interior – then reviewed the data on his helmet display. Three confirmed. He gave the command. "Covering fire! On my mark!"

He led the breach, the fight a brutal, close-quarters affair. Every building was a kill zone, every street a gauntlet. The enemy's discipline was unnerving; they seemed to have an endless supply of fresh troops, and their integrated network meant they knew where to hit, where to flank.

"Ammo check!" Kai gasped, leaning against a crumbling wall, his chest heaving. His M41 was almost empty. The supply lines were choked, the airdrops sporadic and often intercepted. They were running on fumes, on adrenaline, and on the desperate hope that someone, somewhere, was coming.

Air Control In Sector Yankee

Captain Anya "Viper" Sharma, East Asian Airspace

Anya felt the deep, bone-weary fatigue settling in. She had been airborne for what felt like an eternity, landing only for hot refuels and quick re-arms, the cycle of combat relentless. The skies over the East China Sea and the Taiwan Strait were a constant maelstrom of electronic and kinetic engagements. Her F-35A, though robust, was showing the strain. Battle damage indicators blinked across her display – minor hull compromises from shrapnel, some degradation in a sensor array from a near-miss EW blast. The Chinese J-20s, often supported by what intelligence was now calling "Ghost Drones" – small, high-speed, highly stealthy unmanned aerial vehicles that acted as EW platforms and missile decoys – were proving to be formidable adversaries.

"Viper Three-Two, inbound missile, IR signature!" her AI flight assistant warned, its voice calm despite the immediate danger. Anya broke hard, deploying a new generation of smart flares, holographic projections designed to mimic her aircraft's IR signature. The missile, a next-gen variant of the PL-15, tracked the decoy for a terrifying moment before veering off. It was a cat-and-mouse game, constantly evolving.

"EW update: Enemy jamming effectiveness decreasing marginally in Sector Yankee," Anya's lead, callsign "Phantom," reported. "Possible expenditure of high-power EW platforms. Push hard, Viper. We need to maintain some kind of air control over the Strait for the naval assets."

Anya acknowledged, her finger tightening on the stick. The naval battle was brutal. Reports of significant losses for both sides were filtering through, confirming Ellie's intelligence. Without air cover, the ships were vulnerable. Her role was critical, a thin line between survival and oblivion for thousands. She pushed her F-35 towards the most contested sector, the air ahead a shimmering haze of combat.

The Metal Spiders

Specialist Dave Miller, Uijeongbu, South Korea

The first day of the North Korean invasion was a blur of deafening explosions, panicked retreats, and desperate, close-quarters combat. Camp Casey, once a secure hub, was now a smoking ruin. Dave and his surviving squad members were fighting from a hastily established trench line on the outskirts of Uijeongbu, a city just south of the DMZ, serving as the first major defensive choke point for Seoul.

The North Korean artillery continued to pound them relentlessly, accurate and devastating. And then there were the mechanized infantry units – soldiers with lightweight powered assistance frames that allowed them to sprint through shrapnel and climb rubble with disturbing agility. The larger combat automatons, walking tanks of reinforced composite and heavy weaponry, were proving particularly difficult to stop without dedicated anti-armor support.

"Frag out!" Dave yelled, tossing a grenade into a cluster of enemy infantry attempting to flank their position. The explosion sent bodies flying. Their numbers seemed endless. Henderson, the medic, was working frantically, his MOPP suit already smudged with blood and guts. "We've got three critical, Sarge! Need evac, now!" Sergeant Riley, his face a mask of stony determination, shook his head. "No evac, Hendo. Airspace is too hot. We hold here. If we don't, Seoul falls. They know it."

Dave peered through his smart optic, which was now showing static interference, occasionally flashing "COMMUNICATIONS INTERRUPTED." The enemy's EW efforts were intense here, targeting their tactical networks, trying to isolate them. He saw a

North Korean officer, clad in heavier, more advanced armor, issuing commands to a group of soldiers, their voices amplified by integrated comms.

Suddenly, the ground shook with a different kind of impact. Not artillery, but something heavier, something thudding closer. Through a gap in the smoke, Dave saw it – a massive, multi-legged heavy assault platform, a spider-like robot bristling with missile pods and a rotating auto-cannon. It was unlike anything he'd ever seen, clearly a direct import from Russian or Chinese military tech, designed for urban assault.

"Holy mother of God..." Hendo breathed.

"Focus fire! Aim for the joints!" Riley screamed, his voice strained. Dave squeezed the trigger, his rifle rounds pinging uselessly off the automaton's armored plating. This was a nightmare. This wasn't just a conventional invasion. This was a brutal, technologically augmented surge designed to overwhelm and break them. The first day had already redefined the meaning of war.

D + 2

April 20th, 2027

The third day of the war dawned with the realization that this was not a quick strike, but a grinding, brutal conflict, stretching resources and testing the resolve of every combatant from every nation.

A Horrific, Elaborate Game

CTT2 Ellie Rodriguez, USS Ronald Reagan, West Pacific

The Reagan was now operating under extreme EMCON, a ghost in the water, trying to evade a cunning and relentless enemy. The intel center was a dungeon of screens, lit only by the data flowing across them. Ellie's eyes were bloodshot, her mind a numb blur of algorithms and signatures.

"Sir, we're seeing coordinated Iranian EW activity in the Persian Gulf," Ellie reported, her voice raspy. "It's mirroring some of the adaptive patterns we've detected here. They're definitely sharing intelligence, possibly even AI algorithms." Lieutenant Jenkins pounded a fist on his console. "The bastards. So this 'axis' isn't just political. It's a technological fusion."

"That's the assessment, sir. It implies a deeper level of integration than we anticipated. Their EW capabilities are synergistic." Ellie pulled up a new projection. "We're also tracking significant Chinese naval deployments out of the Strait of Malacca – likely heading for the Indian Ocean, possibly targeting our allies' supply lines or even moving to support Iranian operations in the Gulf."

The global chessboard was expanding, each move a ripple across the entire world. The economics of this war were already staggering. Global shipping routes were disrupted, oil prices skyrocketing, and financial markets were in free fall. This wasn't just a military conflict; it was an economic war of attrition, designed to cripple the Western world from multiple angles.

"Keep a close watch on those Chinese movements, Rod. And for God's sake, if you see any new patterns in that Iranian EW, flag it immediately. We need to counter their predictive capabilities." Ellie nodded, her fingers hovering over the touch screen. The invisible war of data and deception was growing more complex, more deadly. Every flicker of a signature, every intercepted byte, held the potential to save or doom thousands. She was tired, but the stakes were too high to falter.

The Despair Of Psychological Warfare

Sergeant Kai Chen, Streets of Taipei

The fighting in Taipei had devolved into a grim stalemate, a bloody siege. The Marines, along with their Taiwan counterparts, were holding a shrinking perimeter around the city center, but they were running low on everything – ammunition, water, and hope. The air tasted of ash and despair.

"Sarge, another surrender broadcast from the Chinese," a Marine called out, the distorted voice of a loudspeaker echoing through the ruins. "They're offering safe passage for civilians if we lay down arms. It's a lie, right?"

Kai didn't answer, just stared at the shimmering heat haze where a Chinese sniper drone was methodically picking off exposed positions. The psychological warfare was as relentless as the kinetic. They knew the enemy would not honor surrender terms for fighting forces, and the "safe passage" for civilians was a cynical ploy to sow discord.

"They're consolidating their heavy automatons on our east flank!" a spotter shouted over the comms. "Two more spider-tanks inbound! They're hitting us with everything!" The heavy assault platforms were the bane of their existence, nearly impervious to their light anti-armor. Kai knew they needed air support, but the skies over Taiwan were still heavily contested, with Chinese air superiority fighters and EW aircraft making bombing runs a suicide mission.

"Get ready to disperse!" Kai barked, eyeing the approaching automatons. "We draw them into the killing zone! Focus fire on the optical sensors! Don't waste rounds!" He watched the spider-tanks advance, their multi-jointed legs scuttling over rubble, their autocannons spitting fire. He thought of Walker, dead on the first day. Matthews, gone. The faces of the fallen, etched in his memory. This was a war of numbers, of overwhelming force, and they were drastically outnumbered. The green hell had become a concrete tomb, and they were fighting on borrowed time far from home.

A World On Fire!

Captain Anya "Viper" Sharma, Kadena Air Base

Anya had just landed from another harrowing combat air patrol, her F-35A covered in sensor-dulling debris from exhaust and atmospheric particulate. The constant high-G maneuvers and EW countermeasures left her physically drained, her muscles screaming in protest. "Intelligence reports confirm significant activity at Chinese space launch facilities," the briefing officer announced, his voice low with despair. "Potential deployment of offensive ASAT capabilities, or deployment of additional EW satellites. They're trying to cripple our space-based comms and navigation."

The news sent a chill down Anya's spine. If they lost GPS and satellite comms entirely, the technological edge of their F-35As would be severely degraded. It would turn their smart fighters into expensive dumb jets, forcing them back to relying on ground-based radar and older navigation methods, severely limiting their effectiveness.

"We also have confirmed Iranian drone attacks on Israeli air defense batteries," another officer added. "Coordinated with a Russian-backed Syrian ground offensive. The Israel front is escalating rapidly."

The world was burning, and Anya was just one pilot in a vast, desperate effort to put out the fires. Her next mission: a deep strike into contested airspace, targeting a newly identified Chinese EW ground station on a remote island. It was a high-risk mission, vital to restoring their critical satellite links.

"Viper Three-Two, you're on the manifest for 'Operation Ripped Net,'" her flight commander said, his face stoic and determined. "Target package is being uploaded. You'll have limited support. Good luck, Captain."

Anya nodded, her stomach a tight knot. Luck wasn't going to be enough. This was a fight not just for the skies, but for the very digital infrastructure that underpinned modern warfare. She walked back to her F-35, strapping in, preparing to gamble her life in the hopes of winning back a sliver of the information war.

Backs Against The Wall As Hope Fades

Specialist Dave Miller, Underground Passage, Uijeongbu

The battle for Uijeongbu was turning into a meat grinder. The North Koreans, bolstered by Russian armor and Chinese combat automatons, were pouring through the breaches in the outer defenses. Dave and his squad were now huddled in an underground passage beneath a collapsed section of a shopping district, acting as a small, desperate rear-guard.

"They're bringing up the heavy flamethrower units!" a soldier screamed, the distinct, terrifying hiss of burning napalm audible even underground. "And the gas! I smell something!" "MOPP level 4! Now!" Sergeant Riley roared, his voice cracking. Dave fumbled with his MOPP mask, the rubber seal pressing uncomfortably against his face, the air smelling faintly of carbon. The threat of chemical weapons had been ever-present, but now it was real.

Through his smart optic, which was now completely offline, Dave could see the blurry forms of North Korean soldiers, their lightweight powered assistance frames allowing them to clear debris with ease. The combat automatons, unaffected by the gas, were methodical in their advance, clearing rooms with precision.

"We're getting cut off!" Hendo yelled, pointing to a flickering screen on his comms unit. "Their EW is completely severing our internal comms! No comms to command, Sarge!"

They were isolated. Alone. The battle for Seoul was descending into urban hell, street by agonizing street, building by building. The sheer, overwhelming force of the North Korean and allied invasion was simply too much. They were being pushed back, their numbers dwindling, their hope flickering.

Dave gripped his rifle, its composite frame cold against his gloved hands. He thought of his brother, safe, he hoped, back home. He thought of the promises of a quick war, of a decisive victory. Those promises had shattered with the dawn. This was a war without end, a brutal, relentless onslaught that demanded everything, and promised nothing but death. He braced himself, listening to the thud of heavy boots approaching their position. Hoping the other battalions had been more successful holding their choke points than his squad had been.

D + 4

April 22nd, 2027

The fifth day was marked by pivotal losses and the harsh realization that the allied forces were now in a desperate struggle to regain the initiative, battered and bruised by the enemy's coordinated precision.

Digital Hide and Seek

CTT2 Ellie Rodriguez, USS Ronald Reagan, West Pacific

The Reagan's intelligence center felt like the nerve center of a dying giant. The ship had taken another hit, this time to a flight deck elevator, impacting sortie generation. Repair crews worked frantically, but the sheer volume of attacks was overwhelming. Ellie stared at a new overlay on Argus: projecting global economic fallout. Supply chains were collapsing, energy grids were under constant cyber attack, and commodity prices were soaring. The political stability of several non-aligned nations was teetering.

"Confirmed loss of the USS Chosin, sir," a junior officer reported, his voice devoid of emotion. The Chosin, a guided-missile cruiser, had been one of their key escorts. "Anti-ship missile strike. Believed to be a salvo from that new submersible platform the Chinese are fielding."

Ellie felt a cold dread. Another major loss. The Chinese "ghost subs" and their long-range missile capabilities were proving to be a formidable threat. Argus was now struggling to process the sheer volume of encrypted enemy communications, their adaptive AI EW systems too complex and somehow rapidly evolving.

"Sir, new intercepts from Iran," Ellie said, her voice strained. "They're celebrating a significant breakthrough in their drone program. Mentions of 'long-range swarm capabilities' and 'autonomous targeting.' This correlates with the increased drone attacks on Israeli defenses." Jenkins slammed his hand on the console again, a raw frustration in his eyes. "They're linking their systems! Their battlefield data, their EW, their targeting... it's all talking. While we're still fighting through bureaucratic firewalls and legacy systems." He looked at Ellie. "We need to break their network, Rod. Find the central node. Find the vulnerability."

Ellie nodded, her fingers hovering over the screens. Breaking their network was like finding a single, shifting grain of sand in a typhoon. But she had to try. The fate of the fleet, perhaps the war itself, hinged on it. The invisible war was turning into a desperate game of digital hide-and-seek, with the survival of nations as the prize.

Fear And Fatigue

Sergeant Kai Chen, Taipei, Taiwan

The last holdouts in Taipei were being systematically crushed. Kai and his handful of remaining Marines, alongside a dwindling force of Taiwan soldiers, were fighting their last stand in the ruins of a historic temple, its ancient stones now scarred by modern warfare. The air was thick with the dust of antiquity and the potent smell of gunpowder and heavy explosives.

"They're deploying humanoid combat automatons now, Sarge!" a young Taiwan Officer shouted, his eyes wide with a mix of fear and disbelief. "They look like soldiers, but they're not!" Through the smoke, Kai saw them. Bipedal robots, eerily human in their gait, but moving with the stiff, purposeful stride of a machine. They were heavily armored, their integrated weapons systems chillingly precise. These weren't the spider-tanks; these were designed for close-quarters urban combat, immune to fear, tireless, and devastatingly effective.

"Focus fire on the joints! Aim for the head-sensors!" Kai screamed, emptying his last magazine into one of the approaching automatons. His rounds ricocheted off its chest plate. The enemy's strategy was clear: overwhelm them with numbers and advanced robotics, designed to bypass human fatigue and fear. The Chinese infantry, aided by their powered load-bearing exoskeletons, advanced behind these automatons, sweeping through the ruins, their adaptive camouflage shimmering, making them almost impossible to track.

"Surrender is your only option!" a translated voice blared from a Chinese loudspeaker drone overhead. "Lay down your arms! Resistance is futile!" Kai looked at the faces of his men, exhausted, bleeding, but still defiant. They were out of ammo, out of water, and out of options. He saw the gleam in the eyes of the approaching humanoid automatons, the mechanical precision of their movements. This was the end of Taipei.

"Fix bayonets!" Kai roared, his voice raw, pulling out his combat knife. It was a futile gesture, a last act of defiance against the inevitable. But it was all they had left. The green hell of Taiwan had become their final resting place.

A Gamble In A Digital Minefield

Captain Anya "Viper" Sharma, Kadena Air Base, Okinawa

Kadena was a mess. The Chinese ballistic missile strikes had been devastating, cratering runways, destroying hangars, and forcing the Air Force to scramble its remaining assets to alternate, smaller airfields across the Pacific. Anya's F-35A was being frantically patched up, but sustained operations from a crippled base were barely feasible.

"We're diverting remaining F-35s to Guam and Australia," the new Ops brief stated, the briefing room half-empty. "Kadena will be limited to emergency landings and minimal sortie generation. Our long-range strike capabilities were severely impacted. And the Chinese have confirmed deployment of their orbital anti-satellite (ASAT) capabilities. We're seeing intermittent GPS degradation and satellite comms outages across the Pacific."

The news was worse than Anya had feared. Losing their space-based advantages was crippling. Her F-35 relied heavily on GPS for precision targeting and navigation, and satellite comms for real-time intelligence and coordinated attacks. It was like fighting with a blindfold and ear muffs.

"Viper Three-Two, you're on the next deep strike," her flight commander said, his face emotionless. "We need to target their orbital control nodes. It's a low-probability-of-return mission. You'll be flying almost entirely on manual navigation, relying on preloaded maps and celestial navigation."

Anya nodded, it was a suicide mission but a necessary one. If they lost space, they lost the war. She walked out to the tarmac, the air tasting of jet fuel and desperation. The F-35 felt different now, less a marvel of technology, more a desperate gamble. She strapped in, her hands gripping the stick, knowing this could be her last flight. The sky, once her sanctuary, was now a digital minefield, and she was flying into its heart.

A Battle Of Body And Will

Specialist Dave Miller, Uijeongbu, South Korea

The underground passage had collapsed behind them, trapping Dave and what few remained of his squad in a desperate, underground fight. The smoke, dust, and the acrid smell of spent chemical agents all around them. His MOPP mask was fogging, and he was struggling to breathe. "Hold this choke point!" Sergeant Riley roared, firing a burst from his rifle. "Don't let them through!"

The North Korean infantry were pushing relentlessly but their movements were constrained by the confined passageway. Behind them, the dreaded heavy assault platform continued to tear through the reinforced concrete, its multi-jointed legs ripping through the structure, trying to create an opening.

"They're using sensory overload weapons!" Hendo screamed, clutching his head, his helmet's internal comms suddenly emitting a high-pitched, debilitating shriek. "It's a sonic blast!"

Dave felt the piercing pain in his ears, even through his helmet. The North Koreans were deploying sophisticated sonic and flash-bang devices, designed to incapacitate their enemies without lethal force, allowing their combat automatons to move in for the kill. His vision blurred, his movements became sluggish. This wasn't just a physical battle; it was a psychological one, designed to break their will.

Through the haze of pain, Dave saw a shadowy figure emerge from the dust – one of the combat automatons, its heavy machine gun whirring. It was methodical, unfeeling, sweeping the corridor with deadly precision. He raised his rifle, but his hands trembled, his body refusing to obey.

Dave knew it was over. They were out of ammo, out of options, and out of time. The steel walls of the underground passage seemed to close in, echoing with the sounds of their last, futile stand. The frozen edge of the DMZ had become a subterranean grave. The war was just beginning, and it was already more brutal, more technologically advanced, and more terrifying than any of them had imagined. Reinforcements from the south could not get here quick enough.

D+5

April 23rd, 2027

The sixth day of the war deepened the resolve of some, while shattering the spirit of others. The battlefields continued to shift, and the global conflict revealed its true, grinding nature.

A Dark Shadow And A Glimmer Of Hope

CTT2 Ellie Rodriguez, USS Ronald Reagan, East of Guam

The Reagan was now limping, not actively steaming, but holding a defensive posture, a casualty of the invisible war. The hit to the flight deck had been worse than initially assessed; combined with the persistent EW attacks, sortie generation was almost at a standstill. Ellie watched the casualty figures tick up on a consolidated display: ships lost, aircraft grounded, personnel wounded or killed. Each number was a fresh cut.

"Sir, we're seeing an unprecedented spike in ransomware attacks targeting civilian infrastructure in the US and Europe," Ellie reported, her voice tight with weariness. "Medical facilities, power grids, financial institutions. It's a coordinated global cyber offensive, clearly linked to the EW patterns we're seeing here. The Iranian-linked proxies are hitting the oil pipelines in the Gulf." Lieutenant Jenkins rubbed his eyes, his face etched with exhaustion. "Economic warfare. They're trying to collapse us from within. What's the latest from the Korean front?"

Ellie swallowed hard. "Seoul is largely contested, sir. Urban fighting. High casualties. DPRK forces are using what appears to be a new generation of autonomous infantry drones but the ROK forces are moving from the south to aid in holding the line. They're devastating in close quarters. And the Chinese... they've established multiple lodgments on Taiwan's Coast. They're attempting to cement their control."

The news was a gut punch. Taiwan, effectively fallen. Seoul, at risk on the front line. The vastness of the Pacific felt less like a buffer and more like an endless, vulnerable expanse. The Reagan, once the spearhead, now felt like a beached whale, its teeth blunted by an enemy who fought on multiple, interconnected fronts.

"We've identified a potential weakness in their primary quantum-encrypted network," Ellie continued, pointing to a shimmering cluster of algorithms on her screen. "It's a tiny, almost imperceptible tremor, something their adaptive AI isn't correcting for. If we could hit it... a full-spectrum cyber-kinetic counter strike might be possible."

Jenkins looked at the screen, then at Ellie. "A shot in the dark, Rod. One shot. But it's all we've got. Send that data up the chain. Immediately. This could be our only way to get back into the fight." Ellie nodded, a flicker of renewed purpose in her bloodshot eyes. It was a desperate hope, a single thread in a tapestry of despair, but it was a chance.

Starvation And Scorched Earth

Sergeant Kai Chen, Taipei, Taiwan

The last stand at the temple had been a slaughter. Kai didn't know how he was still alive. He and three other Marines, along with two Taiwan soldiers, had slipped through the net of humanoid automatons and armored infantry, now ghosts in their own battlefield. They were deep behind enemy lines, foraging for food and water, evading patrols, the city a death trap.

The occupation was brutal. Chinese military presence was everywhere, consolidating control, establishing checkpoints, hunting down resistance. The humanoid combat automatons patrolled silently, their optical sensors sweeping the ruins, their integrated target acquisition systems chillingly efficient. They moved with an unfeeling precision that made them far more terrifying than any human soldier.

"Sarge, food," a young Marine whispered, holding up a small, bruised fruit found in a deserted market stall. Their rations were gone, replaced by whatever meager sustenance they could scavenge. Kai nodded, chewing slowly, the bitterness of the fruit matching their current circumstances. They were still alive but always being hunted. Their comms were completely jammed, relying on sporadic, encrypted burst transmissions from the resistance networks, if they could even find them.

They witnessed the horrifying efficiency of the Chinese forces: drone swarms conducting sweeps of residential areas, the rhythmic thud of precision strikes, the grim sight of long convoys of civilian vehicles being repurposed for military use, or worse, for moving out civilians to unknown "re-education" camps.

"We need to link up with the resistance," the Taiwan soldier, Corporal Lin, said. "My family... they would be here. We have to find them. Fight for them." Kai looked at the ruined city, the smoke-choked sky. He was a Marine, trained for beach landings and frontal assaults, not urban guerrilla warfare against a technologically superior, relentless enemy. But survival was the new mission. And defiance, a burning ember in his chest.

A Capital Under Siege

Specialist Dave Miller, Seoul, South Korea

The underground passages beneath Uijeongbu had saved them, barely. Dave, Hendo, and Sergeant Riley, along with a handful of other survivors, had managed to escape the collapse, emerging into the shattered, burning outskirts of Seoul. The city was a war zone, a landscape of rubble and fire. "We need to find the South Korean resistance," Riley growled, his face covered with soot. "If we're going to make a stand, it's gotta be here." The North Korean offensive had swept through Uijeongbu with terrifying speed, consolidating their hold. Now, Seoul itself is under attack. The familiar skyline was obscured by plumes of black smoke, the sounds of distant explosions constant.

"Civilians," Hendo whispered, pointing to a small group of Koreans huddled in a bombed-out subway station, their faces pale with terror. "They need help" Dave felt a pang of despair "but what can we do?" There were too many civilians, too much chaos. His M-LOK rifle was now just a heavy piece of metal, its smart optic useless due to persistent EW jamming. They were fighting like ghosts, reliant on their instincts and the sheer will to survive.

Suddenly, a distant explosion, larger than the rest, shook the ground. A pillar of smoke rose from the city center, eerily familiar in its shape. "That's... the National Assembly building," Riley murmured, his voice hollow. "They're hitting political targets. Where are our reinforcements!" The fall of Seoul felt imminent. The North Koreans, backed by Russian heavy artillery and Chinese reconnaissance drones, were pushing relentlessly. They were outmatched, outgunned, and out-teched. Dave looked at the faces of his comrades, their eyes reflecting the same desperate understanding. This wasn't a defense anymore; it was a retreat, a fighting withdrawal from a war they were losing. The frozen edge had turned into a burning heart.

D+13

April 30th: A Promise Delivered

US Submarines had started to make their impact against the less technologically advanced PLAN force. This was the key to regaining control. The US President made a speech declaring many victories in the western Pacific. "Our Marines are helping the Taiwan military hold the line! Our Navy is sinking China's ten to one and our ability to strike the Luzon Strait from the protected Philippine Bases will keep the Enemy inside the South Asian Sea!" An executive order declared the new name to replace adversary nations from international waters. "Our adversaries have shown no regard for international laws of war and have committed crimes against humanity. For this reason I have approved actions that will turn the tide in our favor. With the aid of our allies, we will bring the fight to them and with this official declaration from Congress. We can now say what we have known for far too long. We are at War!" The American people erupted in cheers and fear for they had been at war before but a declaration had not been made for nearly 100 years. This was the official start of a Third World War.

The Joint Nation Alliance that had been a sleeping giant started to spread its wings. The American Navy and Air force had been approved to attack critical targets on mainland China. This was a massive gamble with the lives of our pilots and crew and an escalation that could be provoking or taken as an off ramp. Strike fighters blazed into China's air defenses knowing that they had little chance of survival but if effective could be a major tipping point in the war.

Australia's moved forces into the Philippines to reinforce the island that has been under assault by missile strikes since the start of the war. They would be the replacements for the many US Air assets starting to take flight for their massive assault.

The main land air defense systems were the highest priority targets. For if they could own the sky they could change the tide in an immense way for the allies. The assault was effective but not without cost. A combined action from aircraft carriers pulsing into battle, land based fighters from the Philippines and Guam with tanker escorts. The assault was timed with other assaults in Korea and the mission was complete 90% of targets hit. A major blow to China's Air defense but before they were hit, a rapid air defense drone scramble had emerged, they took out many of our fighters.

The first week of battle had been a blur of sustained, unyielding terror. By the end of the month, the initial shock had curdled into a grim, exhausting reality. The world was at war, irrevocably and undeniably, and the Pacific had become its primary, burning crucible.

The Burden Of The Observer

CTT2 Ellie Rodriguez, USS Ronald Reagan, East of Guam

The Reagan's intelligence center was a mess. All around them was the metallic tang of fear and the acrid scent of ozone. Ellie's eyes, red-rimmed and hollow, darted across the displays that shimmered with the chaotic ballet of war. The "First Pacific Clash" wasn't a historical event; it was a living, screaming monster. She'd seen the blips vanish, felt the concussion of distant impacts that rattled the entire carrier, even at range.

"Another one, Argus," a breathless voice reported over the comms, a new blip, a Chinese destroyer, winking out of existence on a secondary screen. "Looks like a sub-launched torpedo. Bravo section's running analysis."

Ellie nodded numbly, her fingers flying across her console. "Understood. Argus, re-calibrating predictive models for evasive maneuvers. Send updated threat profiles to all combatants." The AI, though battered and struggling, was still learning, still fighting. They were witnessing naval engagements on a scale not seen since Midway, and the sheer volume of data, the relentless pace of attack and counter-attack, threatened to overwhelm even Argus. The clash was consuming their lives, a vortex of missile trails and burning oil that painted the vast ocean in shades of death. They were limping out of the Western Pacific, but the battles felt terrifyingly close.

The sheer scale of it, the countless friendly and enemy contacts flickering across the screen, the intelligence reports flowing in – each one detailing another engagement, another loss, another frantic re-assessment of their enemy's capabilities – it was a maelstrom. Every shift was a gauntlet, every moment a fight for critical data, a struggle to unravel the enemy's intentions from the overwhelming noise of combat. The weight of every lost ship, every lost crew, pressed down on her, a silent, heavy shroud. Every adversary ship sunk was a victory but this did not equate. It just added to the loss. There was no joy in the heart of war.

A Win Inside The Chaos

Sergeant Kai Chen, Near Taipei, Taiwan

The Australians took the mainland assault as an opportunity to help take out major invasion forces in Taiwan and target the large automaton armies in the cover of the chaos. On Taiwan's shattered coastline, the relentless Chinese assault had, by this valiant act, been largely pushed back to small lodgements on the coast. But "pushed back and repelled" were words whispered through bloody teeth and strained lungs. Those lodgments are still at strength and forces on Taiwan are heavily depleted.

Kai, his uniform caked in mud and his face streaked with grime, pressed himself against the sandbagged wall of a makeshift trench, the constant whistle of incoming artillery a soundtrack to his waking nightmare. The amphibious landings, though fierce, had been broken. The airborne incursions, met with a fury Kai hadn't known his countrymen possessed. Taiwan forces, bolstered by years of preparation and real-time U.S. intelligence feeds, managed to repel the last Chinese amphibious landings. The fighting on the beaches and in coastal cities was brutal. But the price. The cost.

He watched a medevac chopper, its rotors beating a mournful rhythm, lift off from a makeshift landing zone, carrying away another broken body, another life irrevocably changed. The beach, once pristine, was now a charnel house of twisted metal, shattered concrete, and the lingering stench of cordite. They had held the line, yes, but China was not done. The naval blockade was a tightening noose, and the endless barrages of missiles, launched from beyond the horizon, spoke of a patience that bordered on terrifying. Every lull in the fighting was just that – a lull, a moment to brace for the next wave, the next push. Kai clutched his rifle, its weight a familiar comfort, and watched the gray, tumultuous sea, knowing the next act of the brutal play was just beginning. Initial Chinese objectives for a swift conquest were frustrated.

A Deep Strike To Turn The Tides

Captain Anya "Viper" Sharma, Kadena Air Base

Kadena was no longer the buzzing hub of precision and routine. It was a scarred, limping beast, barely functioning. Anya, her F-35A now with several more patched holes and a persistent hydraulics warning, felt the weariness deep in her bones. The skies over the Pacific were a free-for-all, a brutal ballet of dogfights and missile avoidance. She'd seen too many friendly blips vanish, too many desperate calls for help swallowed by the static of enemy jamming.

"Another run over the Taiwan Strait, Sir?" Anya asked. Her squadron leader's voice cracked a little. "Not today, today we prepare for a new mission. One much more intricate and time sensitive." He escorted her to a makeshift mobile SCIF and proceeded to go over the details. "Today we will be coordinating an attack on North Korea. This is a top secret mission. They don't know this yet but Japan has decided to turn to the offensive and enter the war! Their leader stated that if they don't help now the missiles will continue to strike them and they could be the next one invaded. The Global Resilience Pact (GRP) was formed when the US, Australia, Japan, Israel and many European NATO countries joined together to prevent this war from spilling further into peaceful nations. The Coalition of Sovereign Powers (CSP) as they were now called; China, Russia, Iran, North Korea, and their aligned states.

She strapped into her cockpit, the familiar embrace of the G-suit offering little comfort against the gnawing fear in her gut. She'd heard the whispers, seen the hurried, stone faced meetings of command. Russia was pushing deep into Ukraine , and the Middle East was a powder keg. Their F-35s were being diverted, patched up and sent to the most critical zones, stretching their already thin resources to breaking point and now this. This mission would take out many of the strongholds by the DMZ but she was on the tip of the spear. Her mission was to be one of eight aircraft to take a deep strike into Pyongyang! If luck is on her side she will hit her target and get out prior to being detected. The North Koreans had mostly dated tech other than the newly supplemented Chinese augmentations. Hopefully they didn't get too advanced.

The deep strike will aid the southern reinforcements to push back the CSP to the old DMZ and reclaim Seoul. However this will also serve as a distraction to hit them where it hurts while they are spread thin. This mission would have a few F-35s cover three helos with SEAL Teams to drop them in the outskirts of Pyongyang. They will be on their own to take out high ranking officers in order to disrupt their chain of command. The F-35 will strike critical networks and infrastructure and escort the helos back.

The "First Pacific Clash" felt less like a battle and more like a grinder, chewing up machines and men with an insatiable appetite. But this was their time to push back! She engaged the thrusters, the jet screaming down the damaged runway, leaving the wreckage of hangars and the lingering smell of jet fuel behind. Every sortie was a gamble, every landing a minor miracle, as the enemy's evolving jamming techniques constantly tested their limits but this time it would be a miracle if successful.

Death From Above!

Specialist Dave Miller, Seoul, South Korean

By the end of April, Specialist David Miller's unit found themselves in a hastily established trench line on the outskirts of Uijeongbu. The North Korean artillery continued to pound them relentlessly, accurate and devastating. And then there were the mechanized infantry units that could sprint through shrapnel and climb rubble with disturbing agility. The larger combat automatons, walking tanks of reinforced composite and heavy weaponry, were proving particularly difficult to stop without dedicated anti-armor support. We needed air support and fast!

"That's a confirmed heavy assault platform, Miller," his sergeant barked, pointing to a monstrous, spider-like robot bristling with missile pods and a rotating autocannon. "Coordinates sent. Requesting everything we've got."

Dave nodded, his fingers expertly manipulating the targeting system, even as his smart optic flickered with static interference. He felt desperate satisfaction as the coordinates locked. The thudding approach of the automaton was unnerving, its mechanical precision terrifying. For every success, there were a dozen setbacks. Reports filtered back of Russian probes along the Polish and Romanian borders, testing the very fabric of NATO's resolve. This wasn't a surgical strike; it was a brute force assault, a war of attrition where every inch of ground was paid for in blood. The biting wind carried the scent of smoke and something else, something metallic and sickening – the smell of human despair. He knew that somewhere, U.S. cyber teams were launching counterattacks, but here, on the ground, it was still a brutal, analogue war, fought with mud, blood, and cold steel. And just as he was giving up on any support to attack the spider. He saw a massive explosion and then a loud crack like thunder.

They yelled and cheered as a F-35 flew over them. This spark of hope may be enough to keep pushing Dave though. But the Southern Reinforcements should have been here days ago. He turned on the comms equipment and peeked over the sandbag walls. Nothing but scorched earth and despair as far as he could see. His heart sank in his chest with the realization that this may be all we have. And just as he let himself sink into that harsh reality he heard it. "Bravo Company air support cleared the way!" The radios were full of static but he knew this was them. He jumped on the comms "This is Specialist Miller 2nd Infantry Division come in Bravo Company!"

Across the Globe, the Shadow War Continued

In secure, windowless rooms far from the physical battles, U.S. cyber teams, many of them running on adrenaline and caffeine, fought a different kind of war. The initial shock of April 17th's crippling cyber attacks had given way to a desperate, furious counteroffensive. They clawed back, lost ground, re-establishing critical digital infrastructure, and then, with fierce determination, unleashed their own digital furies upon Chinese networks. It was a constant dance, a digital pas de deux of infiltration and counter-infiltration, the fate of supply lines and command structures hanging in the balance of algorithms and code. Every successful counterattack was a small, vital victory, allowing precision airstrikes to hit their targets, allowing vital communications to get through. But the enemy was vast, resilient, and shockingly adaptive. The cyber war, a silent, pervasive hum beneath the roar of conventional combat, was everywhere, a constant reminder that even in the absence of direct engagement, the world was still bleeding.

At the end of April 2027, no one celebrated. The brief, desperate peace of the holiday season was a distant, forgotten dream. The world was a canvas of fire, smoke, and digital static. Humanity was simply trying to survive the dawn.

May 2027

Plans Never Live Past First Contact

The naval conflict in the Pacific continues to escalate, evolving into a protracted and attritional battle. Both U.S. and Chinese fleets suffer significant losses, with the engagement zones widening across the South Asia Sea and into the East Asia Sea. Submarine warfare becomes increasingly critical and lethal. The sheer scale of debris, oil slicks, and nuclear reactive water began to severely impact marine ecosystems in the region.

Russia continues to exert pressure on Ukraine, but faces stiffer, more organized resistance. NATO, while avoiding direct military intervention in Ukraine, activates more rapid response forces and conducts large-scale defensive exercises along its eastern flank, particularly in the Baltic states and Poland. Intelligence agencies report increased Russian hybrid warfare tactics, including disinformation campaigns and cyber probes against NATO infrastructure.

Seemingly endless regional escalation in the Middle East. The cycle of attack and retaliation between Iran-backed militias and

Israel intensifies, threatening to draw in other regional powers. Saudi Arabia and other Gulf states increase their air defenses and intelligence sharing with the U.S. as fears of wider regional conflict grow. Covert operations and special forces engagements become more frequent.

Cyber war grinds continue on. The digital conflict becomes a war of attrition. U.S. and allied cyber teams make some gains in disrupting Chinese networks, but Beijing's adaptive AI-driven defenses prove incredibly resilient. Both sides experience significant cyber outages affecting not just military but also civilian infrastructure, leading to widespread disruptions in finance, logistics, and communications globally.

Taiwan's resilience was tested. Taiwan forces continue their dogged defense,suffering heavy casualties on both sides. However, China consolidates air and naval superiority around the island, effectively tightening the naval blockade and launching relentless missile and air strikes, focusing on degrading Taiwan airfields, command centers, and industrial capacity. Civilian casualties mount significantly.

June 2027

The Pacific Reaches A Bloody Stalemate

Neither side achieves decisive victory, but both are heavily depleted. Emphasis shifts from large-scale fleet engagements to smaller, more agile naval and air-to-air skirmishes, drone warfare, and special operations raids. Shipping lanes in the region are virtually shut down, causing severe global supply chain disruptions. The City of Seoul was successfully defended and the front line pushed north of the DMZ. The threat has shifted to the adversary in Pyongyang. Media cycles split between applauding the victory or calling us the new aggressor in the war.

Russia's offensive in Ukraine slows significantly due to fierce Ukrainian resistance, NATO's indirect support, and the impact of U.S. precision strikes on their logistics. The front lines stabilize into a brutal war of attrition, marked by heavy artillery duels and trench warfare, reminiscent of older conflicts. Both sides dig in, preparing for a long haul. The only movements beyond the lines are drone strikes that are becoming more well defended.

While direct engagements continue, the Middle East conflict begins to shift into a more complex geopolitical struggle. Efforts by regional and international actors to mediate or deescalate gain some traction a midst the wider global chaos, though sporadic clashes persist. The focus shifts to securing key oil infrastructure and preventing wider proliferation. The Media is spinning the US involvement into a new desert storm starting a war with no end in sight. Is this a nation building mission or an attempt for regime change and will either last. What is the measure of success? Why should the American people care enough to fight?

The global cyber war becomes a permanent, pervasive feature of the conflict. While major financial markets have implemented emergency protocols, the constant barrage of attacks continues to destabilize economies worldwide. Critical infrastructure in numerous countries experiences intermittent blackouts and service disruptions, impacting daily life and straining emergency services.

Despite their continued resistance, Taiwan finds itself increasingly isolated. The blockade is near total, and the flow of external aid is severely curtailed. Chinese forces begin to make incremental but costly gains on the ground, pushing further inland. The world watches, horrified, as Taiwan's long, desperate fight for survival continues.

July 2027

The Scorching Ultimatum

The blazing heat of July was a physical manifestation of the global temperature. The world wasn't just holding its breath; it was suffocating. Every news broadcast, every hushed conversation among military personnel, every flickering screen in Ellie Rodriguez's intelligence center, was dominated by two terrifying phrases: ultimatum and high alert.

Propaganda, Provocation, And Starvation

CTT2 Ellie Rodriguez, USS Ronald Reagan, Pearl Harbor, HI

Ellie felt the tension in Argus like a tangible weight. The hum of electronics, now vibrated with a brittle silence, punctuated by the sharp clicks of keyboards and the hushed commands of officers. Just days into July, the official statement from Beijing had dropped like a thermonuclear bomb: China's Ultimatum. The map of the Pacific glowed ominously, the red zones around Taiwan throbbing. "Any further U.S. military action in our sovereign territories will be met with full-scale war," the voice, synthesized and chillingly devoid of emotion, echoed through the intel center from a looped propaganda broadcast. "This includes the continued illegal presence of foreign naval assets and aerial intrusions."

A shiver traced down Ellie's spine. Full-scale war. As if what they were already embroiled in wasn't. They were still reeling from the attrition of the "First Pacific Clash," the Reagan's hull patched in a dozen places, her air wing thinned. The implied threat was clear: step one inch further, and the gloves were truly off. The unspoken question was, how much more off could they be? The market data flickering on her secondary screen was an apocalyptic landscape of red. Global markets hadn't just crashed; they had disintegrated. News feeds spoke of supply chains having not merely fractured, but dissolved. The world was starving, literally, and the war machine was still churning. She saw the stress lines on Captain Reynolds' face, deeper than ever. They were already stretched thin, running on fumes, and Beijing was pushing them to the absolute breaking point.

Hopes Of A Weary Warrior

Sergeant Kai Chen, Coastal Taiwan

The Taiwan trenches were consumed with humidity and the smell of fear. Kai Chen, now leaner, harder, his eyes holding a haunted wisdom, watched the relentless Chinese bombardment from his position. The "ultimatum" had reached them even here, through whispered radio broadcasts and grim-faced commanders. It felt like a cruel joke. They were already fighting for their lives, daily, hourly, against an enemy that seemed to have an endless supply of men and materiel.

The Chinese forces, though frustrated in their initial rapid conquest, had not let up. The naval blockade tightened daily, squeezing the island's last lifelines. And with the global economic collapse, the flow of vital supplies from outside had slowed to a trickle, then to nothing. Kai could see it in the faces of his comrades – the growing hunger, the dwindling ammunition. He heard the commanders talking in hushed tones about China's Gamble: a "final push" for Taiwan. Beijing, facing internal economic turmoil from the sanctions and the war, seemed determined to secure its prize, whatever the cost. Kai knew what that meant. It meant more waves, more bombs, more blood on the beaches. He gripped his rifle, its worn stock a cold comfort. They had repelled them once. Could they do it again? Hope felt like a dangerous luxury.

From One Front To The Next

Captain Anya "Viper" Sharma, Relocated to Guam Air Base

Despite a successful mission in Pyongyang and a push further north, the shift from Kadena to Guam had to happen. It was a stark reminder of their vulnerability. Kadena was still operating, but barely, under constant missile threat. Guam, a vital rear-area hub, was now crammed with battered air-frames and exhausted crews. Anya's F-35A had survived, but it felt more like a stubborn, protesting old workhorse than a sleek war machine. The news from Europe was chilling. Russia's Nuclear Threat. Moscow had placed its nuclear forces on high alert, demanding NATO withdrawal from Eastern Europe.

Anya watched the grim faces of the intel officers poring over satellite imagery. The implications were clear: NATO was caught in an impossible bind. NATO's Dilemma was being played out in real-time debates across European capitals, broadcast on the sparse, government-controlled news feeds. Direct intervention in Ukraine risked nuclear annihilation. Inaction risked the collapse of the European security order. Anya, staring at the scarred tarmac of Andersen Air Force Base, felt a cold knot in her stomach. It wasn't just about dogfights and missile locks anymore. The stakes had spiraled beyond anything she could comprehend. The thought that one wrong move, one misplaced word, could unleash hell, was a constant, suffocating presence. Anya's Commander came up to here and said "I know you just got here but we need you in Poland. The war is shifting and this is where your skills come in. You will need to take a few hops to get there but we will be sending you out by the end of the week" Anya sighed knowing that she just traded one front of the war that was settling for a newer, edgier one.

The Ping Pong Of Fallen Capitals

Specialist Dave Miller, North Korean Border, South Korea

The roar of the advancing South Korean armor was a welcome, if terrifying, sound. After weeks of grinding, bloody combat, they had done it. North Korea's Capital Falls. Pyongyang, once an impenetrable fortress, was now in allied hands. Dave, covered in sweat and grime, watched as malnourished civilians, gaunt and broken by decades of oppression, cautiously approached their lines, some praising the new leadership from the South, others simply begging for food and water. They migrated south in a slow, mournful tide, their faces etched with suffering.

But the war wasn't over. The front line had been pushed to the northern mountains, a jagged, unforgiving landscape. And even some of the stronger, younger North Koreans, fueled by a visceral hatred for their former regime, were joining the South Korean forces, picking up rifles and turning them on their previous oppressors. It was a bizarre, unsettling sight. Dave, scanning the tree line for hidden ambushes, knew this was a new kind of fight, a brutal insurgency in the mountains.

The victories, when they came, felt hollow, overshadowed by the greater, looming dread. He heard snippets of news from overseas – the economic collapse, China's ultimatum, Russia's nuclear threats. The war in Korea might be turning, but the world was still teetering on the precipice, and the true nightmare felt like it was only just beginning. As July bled into August, the heat intensified, both physically and metaphorically. The world held its breath, suspended in a moment of terrifying indecision. The war machine was already in motion, and no one knew how to stop it.

August 2027

The Unseen Pressure

August arrived, not with the promise of summer's ease, but with the suffocating weight of continued crisis. The ultimatum and threats not withdrawn, only intensified by the silence of a world paralyzed by fear. The war machine, fueled desperation and a chilling inevitability, continued its inexorable grind.

Global Economic Collapse

CTT2 Ellie Rodriguez, USS Ronald Reagan, Pearl Harbor, HI

On the Argus, Ellie tracked the harsh statistics: more ships lost, more pilots missing. The Reagan, a floating city of steel, was now navigating a true minefield of conflicting orders and near-impossible demands. Beijing's ultimatum from July hung over every decision, every movement of the fleet. "Maintain deterrence without provoking direct escalation," the new directive read, a phrase that seemed to contradict itself.

The global economic collapse was now painfully tangible. Rations were tighter, and the occasional incoming supply ship was celebrated like a victory. News of riots in port cities, desperate populations scrambling for dwindling resources, filtered through their secure channels. Argus was now devoting significant processing power to tracking global supply chain disruptions, not just enemy movements. The war had metastasized beyond the battlefield, infecting the very sinews of civilization. Ellie found herself watching the red lines of economic graphs plummet with almost as much dread as the red blips of enemy contacts. The sheer, quiet desperation emanating from those distant economic centers was a chilling counterpoint to the mayhem in the Pacific.

Enduring Struggle Of A Starving Nation

Sergeant Kai Chen, Coastal Taiwan

August in Taiwan was a cruel furnace. The China's Gamble was playing out in slow motion, a relentless tightening of the noose. The constant bombardments were now accompanied by an increase in probing attacks, not full-scale invasions, but relentless tests of their defenses, wearing down the already exhausted Taiwan forces. Food and medical supplies were critically low. The wounded lay in makeshift clinics, their chances of survival dwindling with each passing day. The blockade has effectively starved the nation.

Kai watched as a small, civilian fishing boat, clearly attempting to break the blockade, was vaporized by a precision missile strike far out at sea. The message was unequivocal. Beijing was utterly determined to starve Taiwan into submission if direct invasion continued to prove too costly. The faces of his comrades were gaunt, their eyes hollow, but the stubborn fire of defiance still flickered within them. They knew this was the endgame. Either they broke, or China did. But the idea of China breaking seemed as likely as the sky falling. Every day was a question of how much more they could endure.

A Pawn In A Larger Ploy

Captain Anya "Viper" Sharma, Poland, Eastern Europe

The Russia's Nuclear Threat from July had not dissipated; it had become a terrifying, constant lingering beneath the surface of all strategic discussions. Anya, back from another grueling patrol, debriefed her mission, the F-35's combat data uploaded for analysis. Her flights now often involved skirting the fringes of sensitive airspace, testing response times, witnessing firsthand the sheer tension on the European front.

The NATO dilemma was now an open, agonizing wound. European nations were debating direct intervention against Russia with an intensity that verged on desperation. She saw the news reports, the fiery speeches from parliamentary leaders, the stoic generals. The risk of nuclear escalation was terrifyingly real, yet the alternative, allowing Russia to further destabilize Europe, seemed equally catastrophic. Anya felt like a pawn in a game played by giants, her F-35 a tiny, fragile piece on a vast, unforgiving board. She found solace only in the cold, efficient beauty of her aircraft, its systems a familiar comfort in a world gone mad.

Rationing On The Ridge Lines

Specialist Dave Miller, North Korean Mountains

The Korean peninsula was a testament to the brutal August heat and the even more brutal reality of mountain warfare. Dave and his unit were now locked in a grinding, vicious fight for every ridge line and valley. The North Korean forces, though shattered, were proving resilient in their desperate retreat. And then there were the new recruits, former civilians now armed with their eyes burning with a mixture of revenge and fanaticism, fighting alongside the South Korean forces. It was a chaotic, unsettling alliance.

The news from the outside world, when it filtered through, was grim. The global economic collapse meant fewer supplies, less support, and more pressure on their already strained logistics. Dave saw the effects firsthand: the meager rations, the worn-out equipment that couldn't be easily replaced. The war in Korea might be geographically contained, but its tendrils reached everywhere, strangling resources, exacerbating suffering. He looked at the scarred peaks around them, knowing that this was where the North Korean regime intended to make its last stand. And he, along with the rest of his unit, was stuck in the middle of it. He cleaned his rifle, the familiar ritual, a small anchor in a sea of overwhelming chaos.

As August bled into September, a sense of weary inevitability settled over the world. Diplomatic channels remained open, a desperate scramble for solutions a midst the escalating violence, but the momentum of the war machine felt unstoppable.

September 2027

The Deepening Chill

September brought with it the first biting chill of autumn, a bleak promise of winter's return. The world had passed the point of holding its breath; it was now exhaling slowly, a long, weary sigh of resignation. The scramble for diplomatic solutions continued, a frantic, desperate dance on the edge of a precipice, but the war machine, having gained a terrifying momentum, showed no signs of slowing.

An Endless AI War-Front

CTT2 Ellie Rodriguez, USS Ronald Reagan, Pearl Harbor, HI

Ellie, haunted by the ever-present drone of Argus and the constant stream of bad news, felt the exhaustion gnawing at her bones. Beijing's ultimatum remained, a cold, hard fact, shaping every strategic decision. The economic crisis had deepened, impacting even the mighty U.S. Navy. Maintenance backlogs grew, spare parts became gold, and the taste of recycled water was a constant reminder of dwindling resources. Argus, ever vigilant, was now flagging a disturbing increase in Chinese AI activity, not just probing their defenses, but actively adapting, learning at a horrifying pace. The strategic implications were terrifying. They were not just fighting men and machines; they were fighting an intelligence that learned and evolved.

A Lonely Fight On An Everlasting Assault

Sergeant Kai Chen, Coastal Taiwan

September brought no respite to Taiwan. The "final push" hadn't been a single, dramatic assault, but a grinding, unrelenting offensive. Kai could see the impact of China's gamble in the faces of the weary, starving civilians huddling in the shattered cities, in the dwindling numbers of his own comrades. The Chinese, facing their own economic turmoil, were pressing their advantage with a chilling ruthlessness. Every meter gained was at immense cost, but they kept coming. It reeked of smoke and the acidic tang of chemicals. The news, when it reached them, spoke of a world that was simply watching them die, embroiled in its own larger struggles. Taiwan was becoming a forgotten casualty, a small island caught in a global storm. Kai, his body a map of scars, felt the weight of that abandonment.

Watching A Nation Collapse As Europe and NATO Debate

Captain Anya "Viper" Sharma, Poland, Eastern Europe

The chill of autumn did little to cool the heated debates within NATO. Anya, her F-35 patched and re-patched again, found herself flying more strategic reconnaissance missions, witnessing the terrifying scale of Russia's military buildup in Ukraine. Listening to the news of Russia's Nuclear Threat hanging like a dark cloud over every political and military discussion. Yet, the continuous, aggressive probing by Russian forces along the borders was eroding their resolve.

Anya was called to redeploy to the European front, to do what? She thought. She was stuck here watching, unable to act due to legalities and fear. When she saw the flicker of desperation in the eyes of the European pilots she flew alongside, a shared understanding of the impossible choice facing their leaders. The fate of Europe, perhaps the world, rested on a knife's edge, balanced precariously between defiance and self-preservation.

A Snowy Last Stand

Specialist Dave Miller, Korean Mountains

The relentless, brutal mountain warfare in North Korea continued through September. Dave and his unit were now fighting an enemy that was both desperate and fanatical. The North Korean forces, pushed into their last strongholds in the jagged peaks, fought with a chilling ferocity. And the former civilians, now armed and enraged, were proving to be unpredictable, savage, driven by generations of oppression. Basic medical supplies were scarce, and the wounded sometimes simply waited, stoically, for the inevitable. News filtered through of the larger war, the endless stalemate in the Pacific, the nuclear threats in Europe. It made their own brutal, localized struggle feel almost insignificant, yet no less real. Dave found himself staring at the bleak, snow-dusted peaks, knowing that this was where the North Korean regime would make its last, bloody stand. The world might be scrambling for diplomatic solutions, but here, in the cold, unforgiving mountains, the war machine was a grim, undeniable force, carving out its own brutal end.

As September drew to a close, a palpable sense of exhaustion settled over the fighting forces and the beleaguered global populations. The "war machine" was indeed in motion, having devoured the spring and summer of 2027. The scramble for diplomatic solutions felt increasingly futile, a whisper against a rising tide of violence. The world was sliding into a deep, dark winter, with no clear end in sight.

October 2027

The Unraveling

October arrived not with the crisp, clean air of autumn, but with a choking dust and the bitter taste of defeat. The world, already teetering, finally began its agonizing descent. The war machine, having consumed spring and summer, now gorged itself on the fall of nations.

A Nation's Hope Devoured

CTT2 Ellie Rodriguez, USS Ronald Reagan, Pearl Harbor, HI

The atmosphere in Argus felt like a graveyard of vanished blips and silent comms. Ellie watched, numb with horror, as the final, massive wave of Chinese amphibious assaults surged onto Taiwan's shores. Intel feeds, sporadic and breaking up, showed the inevitable. China's massive landing campaign resulted in the occupation of Taipei. It wasn't a conquest; it was an immolation.

Before the fall of Taipei, a series of pre-planned explosions ripped through the island. "Strategic denial measures," a defeated looking commander muttered, watching as industrial complexes and port facilities on the power grid went dark. The US and Taiwan forces destroyed critical infrastructure on Taiwan to prevent China's use of them. A desperate act of self-mutilation, an attempt to deny the victor their spoils. It sent a chilling message of desperation across the Pacific. Argus registered the EMP bursts, the collapse of communication networks, the final, desperate acts of a dying nation. Ellie felt a cold, hollow ache in her chest. Taiwan was gone. Not conquered, but consumed. And with it, a piece of the world's hope.

Left For Dead, Alone And Unafraid

Sergeant Kai Chen, Occupied Taipei

The stench of smoke and death was the new perfume of Taipei. Kai, his body screaming with exhaustion and his mind a swirling vortex of grief and rage, moved through the rubble-strewn streets like a ghost. He had witnessed the occupation of Taipei firsthand, the unending waves of PLA soldiers, the relentless shelling, the final, desperate, futile resistance. He'd seen the flashes as their own people, following desperate orders, detonated key facilities, turning the very infrastructure they fought to protect into twisted, unusable wreckage. The very ground trembled as vital power grids and chip manufacturing plants became craters, echoing with the final, defiant screams of those trapped within.

Now, the chilling propaganda has begun. Loudspeakers mounted on PLA armored vehicles blared an unsettling message in Mandarin, translated by grim-faced locals. China made an appeal to join China's forces. Vowing there will be no more losses. It was a sick joke, a twist of the knife. Kai watched a squad of PLA soldiers, their faces impassive behind their visors, march past. He saw the gleam of new uniforms, the relentless efficiency. This wasn't just an occupation; it was an absorption. They had lost. Everything they had fought for, every drop of blood spilled, every life sacrificed, seemed to have been for nothing at all. The resistance was shattered, scattered, hunted. The weight of defeat pressed down on him, suffocating.

The Collapse Of Another Capital

Captain Anya "Viper" Sharma, Eastern Europe Airspace

The air over Eastern Europe was a maelstrom of electronic warfare and growing dread. Anya's F-35 shuddered, its sensors momentarily blinded by a particularly vicious Russian jamming burst. The news that filtered back to their forward operating base in Poland was grim beyond measure. Russia pushed into the Ukrainian capital. Kyiv had fallen. And with it, a new, terrifying ultimatum. Russian rhetoric, already incendiary, sharpened to a razor's edge, directly aimed at them. They spoke of "buffer zones" and "historical claims" to territories that included Polish soil. Russia is forcing NATO's hand with rhetoric toward Poland Anya thought.

Anya saw the fear in the faces of the ground crew, in the quiet, tense conversations of the pilots. Direct engagement with Russian forces on Ukrainian soil was one thing; a direct threat to a NATO member was another entirely. NATO's dilemma had intensified into a desperate, existential choice. To respond meant risking nuclear war. To not respond meant the collapse of the alliance, nation by nation. She pushed the throttle forward, her F-35 slicing through the bruised sky. The only comfort was the sheer focus of flight, the instinctual dance of evasion and attack. Below, the lines blurred, the future uncertain, but the next target, the next threat, was all too real.

Victory Evolves Into Deja Vu

Specialist Dave Miller, Korean Mountains, South Korea

The chilling, triumphant news reached them in the freezing, snow-dusted mountains of North Korea: North Korea's supreme leader was killed by a Navy SEAL operation on the east coast of North Korea. A momentary surge of satisfaction rippled through Dave's unit. Using satellite imagery they found him in a compound on the east coast by Chongin. They went in at night submerged to the coast and snuck into the compound. They killed him and four other high officers. The brutal tyrant was gone. But victory here felt hollow, overshadowed by the terrifying implications elsewhere. The Korean War, the first one, had ended in a stalemate, a fragile DMZ. This time, the lessons had been learned, brutally.

The news from command was clear: their mission was shifting. They were no longer just pushing the line; they were defining it. The GRP nations had decided to define a new DMZ and fortify in order to avoid mistakes made in the previous Korean War. The winter would quickly get brutal and this could prove to become a crippling disadvantage if they chose not to dig in.

And so they did, not for offensive positions but for fortifications, deep and permanent. Observation posts were reinforced, minefields laid with precision, and the deployment of automated sentry guns began. This new DMZ wouldn't be a fragile line of barbed wire; it would be a steel curtain.

The influx of refugees from the North, once a hopeful sign, now became a logistical nightmare for the South, as the South Korean government strained under the burden. And then came the even more unsettling news, broadcasted on captured North Korean frequencies: China made an appeal to the North Korean people to join China as a new province. Vowing to hold the line from any more losses. It was a chilling appropriation, a ruthless expansion of power, using the death of a tyrant as a pretext. Many saw this as just a power grab but many in the far north saw this as an easy way to get protection. They feared the south would continue their push north and they did not know what would come of them. Generations of brainwashing left many afraid and they saw China as a friend.

Dave looked at the new, imposing fortifications rising from the frozen earth. The war might be shifting, but the grim work of holding the line, of preparing for the next inevitable conflict, had just begun a chilling reminder of where he had been just six months ago.

Six Months Since The World Plunged Into The Abyss.

The initial shock had long since faded, replaced by a cold, enduring reality. The war had not been quick, nor decisive. It had become a global crucible, a grinding, attritional conflict that stretched across continents, bled economies dry, and shattered lives. The initial territorial gains by the Chinese-Russian-Iranian axis had largely stabilized, but at immense cost. The promised rebalance of power was coming, but it was forged in blood, scarcity, and digital darkness. Global trade had fragmented, supply chains were broken, and rationing was a fact of life in every nation.

Across the Western world, public sentiment was a volatile mix of grief, fear, and simmering resentment. Initial patriotic fervor had largely evaporated, replaced by a grudging acceptance of widespread rationing, continuous cyber-induced blackouts, and the ever-present threat of resource scarcity. Food riots, though quickly suppressed, had become common in European capitals. In the

United States, gas prices had quadrupled, and basic goods were often unavailable. Government propaganda channels worked overtime, blaming enemy cyber attacks and extolling the "resilience" of the nation, but the cracks were showing. Civilian casualties from cyber-induced infrastructure failures were mounting, though largely unacknowledged. The refugee crisis from the active war zones spiraled, placing immense strain on neighboring countries and international aid organizations, many of which were themselves struggling under constant cyber assault and funding shortfalls. The global financial system, teetering on the brink, had largely bifurcated, with a new, state-controlled economic bloc emerging within the 'CSP' nations, immune to Western sanctions, effectively creating a parallel global economy. Protests against prolonged war and economic hardship, often fueled by subtle enemy disinformation campaigns, flared periodically, met with increased domestic security measures.

US President, in a televised address to the nation: "We acknowledge the immense burden our citizens are carrying. This is a fight for the very future of liberty, and such a fight demands sacrifice. We are actively developing new strategies to counter the enemy's unconventional warfare tactics and rebuild our economic resilience." (October 5, 2027)

Chief of the Russian General Staff, in an interview with RT Global: "The West underestimated our resolve, and they certainly underestimated our technological prowess. Their hubris led them to believe their digital superiority was absolute. We have proven them wrong. The new world order is not coming; it is already here." (October 14, 2027)

A leaked intelligence memo, widely circulated online via anonymous channels: "Enemy ASAT capabilities have severely degraded over 60% of allied low-earth orbit communication and navigation satellites. Recovery is estimated to take years, even with accelerated launch schedules. Strategic implications are dire." (October 20, 2027) - Though officially denied, this memo further eroded public trust.

Chinese Premier, addressing the People's Congress: "The path to reunification and regional stability is being paved by the unwavering determination of our armed forces. Those who continue to resist the tide of history will find themselves isolated and broken. Our new economic partnerships are forging a brighter future for the East." (October 2?, 2027)

October had consumed Taiwan and Kyiv, pushed Russia to the brink of nuclear escalation, and transformed the Korean Peninsula back into a fortified borderland. The world was unraveling, piece by agonizing piece, and the cold grip of winter was only just beginning.

November 2027

The Vast Shadow

November settled over the shattered world like a shroud, cold and unforgiving. The brutal conquests of October had not brought peace, only a deeper, more pervasive dread. The war machine, fueled by desperation and a chilling logic, continued its inexorable grind, claiming not just territory, but lives, and the very concept of hope.

Promotion And A New Purpose

CTT1 Ellie Rodriguez, Hardened Command Center, Hawaii

The intel center on the Reagan had become a mausoleum, filled with ghosts and the hum of machines plotting devastation. Ellie's eyes were perpetually burning, the digital glow of Argus reflecting in their weary depths. The fall of Taipei and the strategic denial of Taiwan's infrastructure in October had been a bitter pill, a victory for China that tasted of ashes. The USS Ronald Reagan, after sustaining too much damage and its systems proving too vulnerable to the enemy's adaptive EW, had been pulled back to Bremerton for extensive, long-term refit. Ellie, with her unparalleled expertise in hostile EW patterns and their emerging AI-driven nature, had been promoted and reassigned to a hardened command center within Pearl Harbor.

Now, Beijing's triumphal broadcasts echoed across the region, especially their chilling appeal to the North Korean people. China's appeal to the North Korean people to join China as a new province, vowing to hold the line from any more losses, was a geopolitical earthquake. It wasn't just annexation; it was a defiant declaration of a new world order, a clear message to the West that their traditional spheres of influence were being redrawn, brutally.

Ellie watched the analytics. Argus was now projecting complex scenarios for the consolidation of this new "North Korean Province," including the anticipated brutal suppression of any lingering resistance and the rapid integration of their infrastructure into the Chinese system. The implications for the new DMZ in Korea were stark: an even more formidable, unified enemy.

Her new task was the "Shadow Net" initiative – an audacious attempt to build a resilient, quantum-resistant communications and intelligence network using a distributed mesh of repurposed commercial satellites and deep-sea fiber optic cables, designed to bypass the enemy's orbital and terrestrial jamming. It was painstaking, agonizing work, constantly battling the enemy's adaptive AI, which learned and evolved with terrifying speed. But this gave her purpose, a reason to keep fighting on

A Fiery Green Hell To A Cold Gray Cell

Sergeant Kai Chen, Interrogation Facility, Mainland China

The cold, cinder block cell was a stark contrast to the humid jungles and shattered cities Kai had known. The only constant was the dull ache in his bones and the relentless, disembodied voice that filled the silence. Sergeant Chen's capture had been swift and brutal. After weeks of desperate, isolated resistance in the mountains outside Taipei, his small, starving cell had been encircled by Chinese drones using advanced thermal imaging. The last memory was the flash of a stun grenade, the metallic taste of his own blood, and the boots of the PLA.

Now, he was here. He didn't know where "here" was, only that it was a place of endless questions and methodical deprivation. The interrogators cycled through their demands: unit locations, defense plans, U.S. intelligence protocols. He gave them nothing but his name, rank, and service number, a mantra he repeated until his throat was raw. But it wasn't just the physical torment from abuse and starvation. They played him propaganda videos, clips of Beijing's triumphant speeches, images of Chinese flags flying over Taipei. They even showed him the appeal to the North Koreans. "Your struggle is meaningless, Sergeant," a voice, smooth and cultured, told him repeatedly.

"Your America abandoned its allies. Your cause is dead. Join the new order. There is still a place for you." Each word was a barb, designed to chip away at his resolve, to break the very spirit that had kept him fighting. He knew he was a pawn, a psychological victory for them. He had to resist, even if it was only for the ghosts of his comrades.

Small Gains At An Immense Price

Specialist Dave Miller, Republic of Korea's North Border

The Korean mountains, now dusted with the first heavy snows of November, felt like a tomb. He had been on a reconnaissance patrol, part of the grinding methodical work of fortifying the new border. They were mapping out choke points, laying advanced sensor arrays, ensuring that the mistakes of the previous Korean War would not be repeated. He'd seen the faint thermal signature of the new Chinese forces moving into the northern mountains, solidifying their newly claimed North Korean province, a chillingly efficient occupation.

His last moments were a blur of blinding muzzle flashes and the scream of a close-quarter firefight. An ambush. Not by North Korean remnants, but by well-equipped, highly trained PLA special forces, already embedded deep in the newly claimed territory. He fought until the last bullet, a final, desperate act of defiance against the overwhelming tide. His body lay in the frozen earth, a nameless casualty of a war that refused to end, a testament to the brutal reality of a new border. The new line was holding, for now. But the cost was measured in lives, lives like Specialist David Miller's, sacrificed in the deepening shadow of November 2027.

The chilling news had ripped through his unit. The month ended with the world a darker, more dangerous place. Negotiations, held in secret locations across a fractured world, dragged on for months, punctuated by minor skirmishes and constant, low-level cyber warfare. Finally, on a cold, gray morning in late November 2027, a fragile, deeply unsatisfying armistice was declared. No surrender, no clear victor, just a cessation of overt bloodshed born of utter depletion. Nations had fallen, heroes had died, and the lines of conflict, once drawn by geography, were now carved in blood.

December 2027

The Silence Descends

The war did not end with a decisive victory, no grand push to shatter the enemy, no triumphant flag raising. It ended as it had begun: a brutal, slow bleed. By late 2027, the global conflict had reached a point of mutual exhaustion, an economic and human catastrophe so profound that even the most ideologically committed leaders were forced to concede the unsustainability of continued overt hostilities. The global economy was shattered, societies unraveling under the strain of rationing, widespread famine, and incessant cyber attacks. The "re-balance of power" was achieved, but it was a balance of ruin.

The announcement of the armistice was met with a muted, almost disbelieving silence across the globe. There were no spontaneous celebrations, only weary sighs of relief. A year of constant tension, economic hardship, and pervasive loss had dulled any sense of triumph. Public trust in governments, strained by prolonged rationing, civilian casualties from cyber attacks, and a seemingly endless conflict, was at an all-time low. The "re-balance of power" was now starkly evident: the Western world was economically fractured, its supply chains in tatters, its societies exhausted. The 'CSP' nations, while suffering their own immense losses, had consolidated their new territories and economic blocs. The pervasive cyber war continued, a silent hum beneath the surface of the "peace," making widespread recovery agonizingly slow. Mass migrations had reshaped demographics, and international aid organizations struggled to cope with widespread famine and disease, often hampered by ongoing cyber interference and political blockades. The psychological impact on global populations was profound, a collective trauma of a world that had stared into the abyss and come away irrevocably changed.

The US President , addressing a somber joint session of Congress: "We have reached a necessary, albeit painful, pause in active hostilities. This armistice is not a victory, but a testament to the unimaginable cost of modern conflict. We mourn our fallen, we honor the sacrifices, and we begin the arduous task of rebuilding in a world fundamentally altered. The challenges ahead are immense, and unity, both at home and among our allies, will be paramount." (December 5, 2027)

A joint statement from Beijing, Moscow, and Tehran, read by the Chinese Foreign Minister. "Today marks a new dawn for global governance. The era of unilateral aggression and ideological imposition is over. The cessation of hostilities reflects the triumph of a multi-polar world order, where the legitimate security and developmental interests of all nations are respected. We call for cooperation in building a truly equitable future." (December 7, 2027)

A global financial analyst on a heavily censored, international news feed: "While the shooting has stopped, the economic war continues, perhaps even intensifies. The decoupling of global markets is accelerating. We are not entering a period of peace, but rather a new, dangerous geopolitical winter." (December 12, 2027)

An Invisible War Never Ends

CTT1 Ellie Rodriguez, Hardened Command Center, Hawaii

The silence was the loudest sound. The constant hum of the servers in the intel center remained, but the frantic urgency of the flashing alerts, the constant stream of dire reports, had tapered off. Ellie, huddled over her console, watched the final, official communication scroll across the screen: "Armistice effective 00:01 ZULU, 5 DEC 2027. All kinetic operations ceased."

She felt nothing. No elation, no relief, just a profound, aching emptiness. Her eyes, still perpetually bloodshot, scanned the global map on her main display. The red and blue lines, once symbols of active conflict, now just marked new, grim borders. Taiwan was dark, its network integrated into China's. Korea was the only GRP gain was a mostly united Republic of Korea, optimistic of a more peaceful Korean Peninsula, even if the most northern portion was now part of China. The Middle East was a mosaic of shattered states. Ukraine, a devastated ruin, was firmly under Russian influence.

"It's over, Rod," Lieutenant Commander Jenkins said, his voice hoarse, pulling up a chair beside her. He looked even older, his face gaunt, his uniform hanging loosely on his frame. "It's... over."

"Is it, sir?" Ellie whispered, her gaze fixed on the endless, unseen currents of the cyber landscape. Argus was still active, still flagging anomalies. The adaptive enemy AI hadn't shut down; it had simply shifted its focus, moving from kinetic targeting to economic and informational infiltration. "Their quantum network is still active. Their 'Hive Drones' are still reporting. The economic warfare is intensifying, not stopping. This isn't peace, sir. It's just a different kind of war."

Jenkins sighed, running a hand over his face. "Maybe. But the shooting has stopped. For now." He leaned back, his eyes closing. "You've earned some leave, Rod. Go home. See your family." Ellie looked at her hands, shaking slightly. Go home? To what? A world of rationing, blackouts, and constant fear? A world that would never be the same? The invisible war, the one she had fought from the beginning, was far from over. It had simply gone deeper underground, a cold, persistent battle for the very soul of global society. The silence was not peace; it was the uneasy calm before the next storm.

The Honor And Pain Of Resistance

Sergeant Kai Chen, Interrogation Facility, Mainland China

The interrogator came through the steel cell door and told Kai "Ceasefire. All overt hostilities to end. Maintain posture! There is no reason to resist us now. We will send you home just as soon as you answer a few of our questions." Kai felt no relief. Only a cold, bitter resentment. Taiwan was firmly under Chinese control, its people living under martial law, its resources stripped away. He remembered all of the screams of innocent civilians and the friends lost in battle. He continued to give them nothing as he demanded food to be released.

He thought of Sergeant Riley and Hendo, the two other Marines who had made it to the mountains, who had died weeks ago from untreated infections. "It's over," they whispered, his voice thin, "You lost. It's over." Kai stared a thousand yard stare. Lost in his own mind. He thought of the propaganda blaring from the Chinese-controlled media, declaring peace and prosperity for the "reunified province." Lies! He looked at the tattered remains of his uniform. "No," Kai said, his voice raspy from disuse. "It's not over. We didn't surrender and we're still here." His Interrogator struck him so hard he hit the concrete block wall and fell unconscious. There would be no homecoming for him, no parade. Only the torturous work of resistance, a flickering flame in the overwhelming darkness of war.

Scars Of War And Emptiness

Captain Anya "Viper" Sharma, Poland, Eastern Europe

The news of the armistice brought a stunned silence to the briefing room . No cheers, no celebration, just a collective, bone-deep weariness. Anya, her eyes fixed on the projected global map, watched as the kinetic lines of conflict faded, replaced by stark, static borders.

"It's not a victory," her flight commander said, his voice flat. "It's... a pause. We've held the line, at immense cost. Our supply chains are critical. Our space-based assets are severely degraded. Our allies are reeling. But we are still standing." Anya felt the profound weight of the "cost." The empty chairs in the ready room, the faces of pilots who would never return, the constant threat of the hyper-sonic glide vehicles that had decimated their fleet. Her F-35A was being powered down, its engines cooling, the drone that had become her lifeline now just a quiet, empty machine.

"You're grounded, Captain," her commander told her, his gaze holding hers. "Psych eval. Then indefinite leave. You've earned it and you need it." Anya begrudgingly nodded, but she felt nothing. The adrenaline that had fueled her for so long was draining away, leaving her hollow. The thrill of the sky was gone, replaced by the trauma of what she had witnessed, what she had done. The world had changed, and so had she. She remembered seeing the news reports on base, the fuel protests, the power outages in major cities, the public outcry over wartime sacrifices. It was all real, all a direct consequence.

She walked out onto the tarmac, the frozen winter air heavy around her. The familiar roar of jet engines was absent, replaced by a dull, pervasive hum of generators and repair crews. The sky above, once her domain, now felt vast and alien, no longer a battle-field, but a constant reminder of the unseen weapons that still orbited above, waiting. The war might be over for now, but the scars in the sky, and on her soul, would remain forever.

January 2028

The Cold Dawn Of A Fractured World

January 2028 dawned not with a sense of renewal, but with the pervasive chill of a world caught in a deep, dark winter. The global crucible had forged new realities, leaving behind a scarred, fractured landscape where the lines of conflict, though sometimes invisible, were more potent than ever. There was no formal peace, no grand treaty, only the agonizing, exhausted cessation of overt hostilities, a weary pause before the next inevitable storm.

News Of A Nation Divided In A Shattered World

CTT1 Ellie Rodriguez, Hardened Command Center, Hawaii

The new year brought no relief to Ellie. Her days at the hardened command center were a relentless grind of monitoring, analysis, and a chilling sense of dread. The Reagan was still in dry dock, her future uncertain, a potent symbol of the Pacific fleet's battered state. The Chinese claim over Taiwan was a dark reality, and their aggressive posturing towards the Philippines had intensified. Beijing pushed further into the South Asia Sea, threatening U.S. allies with a deliberate, calculating arrogance. Argus, now fully adapted to the new threat environment, detected increasingly sophisticated Chinese cyber probes against civilian infrastructure in Southeast Asia, testing not just defenses, but the resolve of nations.

But within China, the National Congress of the Chinese Communist Party has been influenced by a new challenger for Chairman of the CMC. A general highly critical of the time and methods used to unify Taiwan. He was the leader of the final push that took the island and he has accused the President of having a soft disposition in combat and delayed victory at the expense of the people. This open criticism of the President could have him jailed or worse but as of now he is free and rallying a nation desperate for strength and stability.

The unification of Korea under the Southern Republic was a small, hard-won victory, but even that was overshadowed by the looming threat of China's new "North Korean Province." The new border was heavily fortified, a testament to the sacrifice, but the pressure was constant. The global collapse continued its relentless downward spiral, causing widespread suffering, even in the insulated world of the military. Food rations were tighter, equipment replacements were slow, and the weight of public desperation, filtering through truncated news feeds, was crushing. Ellie knew the next phase wasn't a question of if, but when. The digital war, the unseen war, continued relentlessly, a low, constant hum beneath the surface, preparing the ground for the next, inevitable eruption.

A Cold Desperate Resilience

Sergeant Kai Chen, Interrogation Facility, Mainland China

January dragged on, each day indistinguishable from the last in the confines of Kai's windowless prison cell. The interrogations had shifted, becoming less about intelligence and more about ideological conversion. They showed him carefully curated news reports: the official annexation of Taiwan, the triumphant parades in Beijing, the stern warnings to any nation daring to challenge China's new hegemony. China now declared no one could stop them and openly denounced the US, depicting America as a paper tiger, consumed by its own internal strife.

They taunted him with news of the Philippines, now caught in China's vice. They spoke of a new "Greater East Asia Co-Prosperity Sphere," a chilling echo of history. The appeals to the North Korean people to join China as a new province were now replaced with footage of "re-education" efforts and "integration" projects. Yet, despite the constant psychological assault, Kai held onto a desperate sliver of defiance. He remembered the fierce, unyielding spirit of his people, the faces of his comrades. He knew their "peace" was a lie, built on suppression and fear. He listened to their boasts, their claims of invincibility, and in his heart, he vowed that he would survive, if only to be a witness.

Indecision and Exhaustion

Captain Anya "Viper" Sharma, Poland, Eastern Europe

The Polish winter deepened its grip in January, painting the landscape in shades of white and gray. Anya's squadron was a ghost in the sky, flying reconnaissance missions along the hardening Russian border. Moscow, still weakened by its lengthy war on Ukraine but desperate for the land it once called its own, continued its aggressive probes. Russian forces, though battered, remained formidable, their intentions clear. Russia continued to test NATO's limits with incursions into Eastern Europe, forcing Anya's squadron into hair-raising intercepts over disputed airspace. The armistice may be in effect in the Pacific but it didn't feel any different in Europe.

The debates within NATO had reached a fever pitch. Some argued for a decisive intervention to push Russia back; others clung to the desperate hope of deescalation, terrified of triggering the nuclear threshold. The global economic collapse had hit Europe hard, exacerbating internal tensions and straining alliances. Anya saw the grim determination on the faces of her European counterparts, a shared understanding that their survival depended on standing firm. The threat of renewed kinetic conflict was a constant, unspoken shadow, a silent countdown to the next phase of a war no one truly knew how to win, or even how to survive.

The Dawn Of A New Year

The year 2027 had been a maelstrom. 2028 dawned with a harsh, deeply fractured normalcy. The world had splintered into two major blocs: the Global Resilience Pact (GRP), a collection of Western democracies and their allies, and the Coalition of Sovereign Powers (CSP), led by China, Russia, and Iran had their new allies and conquered territories subsumed under their banners. The "Gray Peace" was a perpetual state of cold war, punctuated by economic sabotage, digital skirmishes, and the constant, nagging threat of renewed kinetic conflict. Climate change and resource depletion, once perceived by many as pressing global existential threats have been pushed to the back burner by the immediate struggle for survival. Public trust in governments, particularly in the GRP nations, was low, and rationing remained a fact of life. The CSP, meanwhile, projected an image of prosperity and order, enforced by strict control and pervasive censorship. Behind the scenes, China could be on the brink of internal regime change or civil war. The General wants to lead the new China but the old order may be set on making him, as so many others have, disappear.

The "Gray Peace" was about to shatter, not with a bang, but with the quiet, chilling hum of machines preparing for war. The lessons of 2027 had been learned, but the price of learning them had been, and would continue to be, astronomical. And so, the world teetered on the brink. The year of the crucible, 2027, had ended with a prolonged, agonizing whimper, leading to an unsatisfying armistice that merely paused the fighting. But the pause was over. By this quiet January morning, the cold war had already started to fester, the technological arms race was well under way, and the geopolitical divides deepened beyond repair. The "long, gray peace" was about to shatter, not with the overwhelming fury of 2027, but with a calculated, surgical precision, as the unseen war, waged in the depths of fiber optics and the chill of orbital mechanics, prepared to unleash its devastating, conclusive blow.